"Riley builds a compelling case for reinforcing and expanding the longstanding links that connect primatology with anthropology. By weaving her own research interests with both historical and contemporary trajectories in the field, she expertly demonstrates how multi-faceted and informative the human dimensions of primatology can be. Considering the severity of the anthropogenic threats faced by most primates today, Riley's reminder of the anthropological roots of primatology comes at a critical time."

– **Karen B. Strier**, Vilas Research Professor and
Irven DeVore Professor of Anthropology,
University of Wisconsin-Madison, USA

"Essential reading for primatologists young and old, Riley provides the historical anthropological context for the emergence of North American primatology, introducing the key contributions made by pioneers in the field that set the stage for contemporary field primatology as we know it. A pioneer in ethnoprimatology, Riley champions the promise it holds for studying non-human primates in the Anthropocene."

– **Michael Huffman**, Department of Ecology and Social Behavior
at the Primate Research Institute, Kyoto University, Japan

"*The Promise of Contemporary Primatology* provides an insightful and well written journey through the history of primate studies and anthropology, highlighting how both are reconfigured as science, and how human-dominated landscapes intersect with research in the Anthropocene. Reclaiming primatology as anthropology is imperative for the field and for the conservation of species and habitats. A must-read book."

– **Shirley C. Strum**, Professor, Department of Anthropology,
University of California, San Diego, USA, and Director,
Uaso Ngiro Baboon Project, Kenya

"*The Promise of Contemporary Primatology* is a must-read for all students of primatology and anthropology. Erin Riley masterfully traces the evolution of primatology over the years to show the field's integral links to anthropology and the very natural emergence of ethnoprimatology. Some previous studies have touched upon the tension in the relationship between anthropology and primatology; yet by framing this relationship against a historical perspective, Riley provides a much more holistic understanding of the kinship between the two disciplines. Riley successfully spotlights the strength of the interdisciplinarity that increasingly characterizes contemporary primatology, and achieves this in an engaging, accessible writing style that will appeal to the interested citizen as much as to scholars of the disciplines."

– **Sindhu Radhakrishna**, National Institute of Advanced Studies, India

THE PROMISE OF CONTEMPORARY PRIMATOLOGY

This book argues for a contemporary primatology that recognizes humans as integral components in the ecologies of primates. This contemporary primatology uses a broadened theoretical lens and methodological toolkit to study primate behavior and ecology in increasingly anthropogenic contexts and seeks points of intersection and spaces for collaborative exchange across the natural sciences, social sciences, and humanities.

The book begins by exploring the American tradition of anthropology, providing historical and disciplinary context for the emergence of field primatology and how it became a part of this tradition. It then examines how primatology transformed into a field dominated by evolutionary approaches and highlights how the increasingly anthropogenic environments in which primates live present opportunities to understand primate adaptability at work. In doing so, it explores how an extended evolutionary approach can help explain behavioral variation in these contemporary environments. Focus is then given to the ethnoprimatological approach, a contemporary approach that provides a pluralistic framework, drawing from the natural and social sciences and humanities, needed to study human-primate coexistence in the Anthropocene. Finally, the book considers how such a crossing of disciplines can inform primate conservation in the future.

An important interdisciplinary reassessment, this book will be of significant interest to primatologists, biological anthropologists, and scholars of anthropology more generally, as well as evolutionary and conservation biologists.

Erin P. Riley is a Professor in the Department of Anthropology at San Diego State University, USA.

NEW BIOLOGICAL ANTHROPOLOGY

Series Editor: Agustín Fuentes, University of Notre Dame, USA

Emergent Warfare in Our Evolutionary Past
Nam C. Kim and Marc Kissel

The Promise of Contemporary Primatology
Erin P. Riley

For more information about this series, please visit: www.routledge.com/anthropology/series/NBA

THE PROMISE OF CONTEMPORARY PRIMATOLOGY

Erin P. Riley

Routledge
Taylor & Francis Group
NEW YORK AND LONDON

First published 2020
by Routledge
52 Vanderbilt Avenue, New York, NY 10017

and by Routledge
2 Park Square, Milton Park, Abingdon, Oxon, OX14 4RN

Routledge is an imprint of the Taylor & Francis Group, an informa business

Library of Congress Cataloging-in-Publication Data
A catalog record for this title has been requested

ISBN: 978-1-62958-070-8 (hbk)
ISBN: 978-1-62958-071-5 (pbk)
ISBN: 978-1-138-31426-9 (ebk)

Typeset in Bembo
by Newgen Publishing UK

To Millie & Paul Riley

CONTENTS

LIST OF FIGURES AND TABLES

Figures

Tables

FOREWORD

Primatology, the focused study of other primates, is a young field. Emerging from both laboratory and field studies, largely by psychologists, in the first half of the twentieth century it became a formal endeavor incorporating anthropologists and biologists during the 1950s and 1960s. As a field, it is multifaceted, drawing on diverse disciplinary and national traditions. Contemporary primatology developed from the interchanges between relatively distinct schools in Europe, North America and Japan, and their mutual aspiration to achieve insight via the integration of the behavioral, morphological, naturalistic and experimental study of humans' closest relatives. This hybridity and the intellectual vibrancy of primatology were initially, and continue to be, driven by primatological studies situated in and around anthropology.

Primatology generated substantive attention in the 1960s and 1970s due to the desire to gain insight into human evolution via the study of our closest relatives. The early data from a focus on apes and large bodied African monkeys (the baboons) as models to better understand human ancestors shaped scientific and public perceptions of humanity and the other primates. The quest to understand primates in their "natural" environment as models for ancestral humans became a driving force for the field. In the 1970s the results from such primate studies formed the foundation for a sociobiology of human evolution. By the 1980s and 1990s the focus shifted, to an extent, and primatological questions expanded to address broader understandings of ecology and behavior. Moving beyond the goal of better understanding humans, primate studies increasingly played a central role in the quest to unravel the patterns and processes of the evolution of behavior and complex social systems in mammals. The study of primates became a key facet in the study of the evolution of social organization through the lens of behavioral ecology. Such research was central in the development of the socioecological model. By the end of the twentieth century this increased density and depth of study and interest revealed that many if not most primate populations were under moderate to severe threat of significant decline, and in some cases extinction. This impelled many primatologists to invest heavily in conservation research and implementation.

Throughout the core trajectory of primatology, both in the seeking of models for human ancestors and in the understanding of the evolution of complex socio-ecological systems, the dominant paradigm argued for the need to focus on primate groups and populations that were in their "natural" habitats. This dominant ideology assumed that primates in "natural" contexts, those devoid of much direct human impact and living outside of anthropogenic landscapes, gave the highest quality data – they reflected the "real" lives of the primates. In this view, those primate groups and populations who overlapped with human groups, landscapes, and lives were "unnatural," corrupted by human influence. Such a perspective resulted in a concerted attempt to avoid including human activities and impacts in the study of the lives of other primates. This turned out to be a grave mistake. Today, in the Anthropocene, we recognize that humans and other primates are intricately entangled in one another's lives, and that this reality has a deep history. To study the other primates, without including human landscapes, actions and ecologies, is to leave a substantive portion of relevant data off the table, and it has limited our abilities to better understand, model and conserve primate populations.

Today we know that the majority of primate populations are, and have been, in complex ecological, social and historical relationships with human populations. Almost all long-term studies, from baboons, to chimpanzees, to muriquis, to the lemurs, demonstrate that interfaces with humans, in a diverse array of contexts, patterns and intensities, play significant roles in the histories, behavior and ecologies of the other primates. The lives of humans and our closest relatives are intertwined in much more complex and interesting manners than most of twentieth-century primatology ever imagined. This is why primatology of the twenty-first century requires an ethnoprimatology. This is also why anthropological training, theory and insight are so critical to a truly effective and engaged contemporary primatology.

Nearly a decade ago I wrote "Humans are literal and figurative kin to other primates, with whom many of us coexist in diverse social, ecological, symbolic, conflictual, and even hopeful contexts. Anthropogenic action is changing global and local ecologies as fast as, or faster than, we can study them. Ethnoprimatology, the combining of primatological and anthropological practice and the viewing of humans and other primates as living in integrated and shared ecological and social spaces, is becoming an increasingly popular approach to primate studies in the twenty-first century. This approach plays a core linking role between anthropology and primate studies and may enable us to more effectively assess, and better understand, the complex ecologies and potential for sustainability in human–other primate communities."[1] Today, this is truer than I ever imagined, and Dr. Erin

Riley is one of the most prominent scholars developing and exploring this reality. Dr. Riley demonstrates, and advocates for, a primatology that is deeply anthropological, fully ethnoprimatological, theoretically and methodologically robust, and ethically grounded. That is why this book is such a critical contribution to anthropology, primatology and beyond.

In this text, Dr. Riley weaves a stellar narrative of the historical, theoretical, political, and structural processes of a deeply anthropological primatology, illustrating why and how ethnoprimatology is the logical and necessary outcome of what we have learned about primates, humans and our relationships from the last seven or so decades of study. She sets the stage with an intricate narrative of the historical emergence and entanglements of anthropological primatology – the rhizomatic tangle of social theory, ethnography, biological anthropology and the other primates. Riley invites readers to immerse themselves in contemporary primatology and evolutionary theory, the complexities of conservation, and convincingly demonstrates the need for a contemporary primatology that relies on "cross-field and cross-disciplinary exchange and a pluralistic theoretical and methodological grounding."

Riley's arguments are carefully laid out and well supported. Her account is particularly engaging not only due to her painstakingly thorough and intellectually rigorous analyses, but also to the manner in which she conveys her experience, her perspectives and her sincere care for the disciplines of anthropology and primatology, the other primates, and her fellow humans. If the promise of contemporary primatology is to be fulfilled it will be in the simultaneously humanistic, scientific, and intellectually generous manner that Dr. Erin Riley lays out for us here.

Agustín Fuentes
The Edmund P. Joyce C.S.C. Professor of Anthropology,
University of Notre Dame

Note

1 Fuentes, A. (2012) Ethnoprimatology and the anthropology of the human-primate interface *Ann. Rev. Anthropol.* 41:101–17 doi: 10.1146/annurev-anthro-092611-145808

ACKNOWLEDGMENTS

The promise of today's primatology is unmistakably multifold. This book shares my view of its promise. It builds on a piece I wrote for *American Anthropologist* on contemporary primatology in anthropology that my colleague, mentor, and friend, Agustín Fuentes encouraged me to expand into a full-length book manuscript. I will admit that there were a number of occasions during the writing of this book that I asked: Why me? Why this book? I am therefore grateful, first and foremost, to Agustín for believing in me and the perspective I bring, and for his patience and guidance during the process.

The ideas, arguments, and perspectives that I share in this book have been percolating for some time, so I must thank the outstanding mentors who have inspired and supported me along the way. Nancy Krusko and Larry Breitborde, my primary mentors during my undergraduate years, were instrumental in solidifying my interest in primatology and anthropology, and in helping me begin to see how and why the subdisciplinary borders of anthropology could and should be crossed. I am grateful to my PhD advisor, Carolyn Ehardt, for being open to exploring what the ethnoprimatologial approach has to offer, for pushing me to think through and beyond, and for investing in me wholeheartedly. I continue to seek guidance from Carolyn, and she remains as willing and as helpful as ever. I am thankful to Pete Brosius for facilitating my first connections in my now beloved Sulawesi, and for his intellectual and logistical guidance throughout my graduate training. Thanks also go to Irwin Bernstein and Dorothy Fragaszy for their insight and support as dissertation committee members. The need for, and benefits of, mentoring does not cease once a degree is in hand. Shirley Strum, Karen Strier, and Agustín Fuentes have been significant mentors and role models post-Ph.D. I thank them for helping me sharpen my ideas and for sharing their critical insights.

One's scholarship does not develop in a vacuum; rather, we learn osmotically and through direct engagement with others, reflecting, building upon, and extending ideas conceived and arguments made. It is indeed a privilege to be among communities of great thinkers and productive scholars from which to draw intellectual inspiration. I know that my thinking has benefitted greatly from the work of, and/or intellectual exchanges with, a number of people, in particular (listed alphabetically): Irwin Bernstein, Michelle Bezanson, Lara Braff, Pete Brosius, Colin Chapman, Colleen O'Brien Cherry, Loretta Cormier, Phyllis Dolhinow, Kerry Dore, Carolyn Ehardt, Joe Erwin, Linda Fedigan, Agustín Fuentes, Paul Garber, Donna Haraway, Catherine Hill, Katie Hinde, Kimberley Hockings, Michael Huffman, Katharine Jack, Iskandar Kamaruddin, Barbara King, Eben Kirksey, John Knight, Susan Lappan, James Loudon, Katherine C. MacKinnon, Nicholas Malone, Jonathan Marks, Andrew J. Marshall, Matthew McLennan, Robin Nelson, Sindhu Radhakrishna, Amanda Rees, Melissa Remis, Carolyn Jost Robinson, Jessica Rothman, Julienne Rutherford, Joanna Setchell, Christopher Shaffer, Anindya Sinha, Barbara Smuts, Charles Southwick, Leslie Sponsel, Karen Strier, Shirley Strum, Robert Sussman, Chia Tan, Bernardo Urbani, Erin Vogel, Paige West, Bruce Wheatley, Rebecca Witter, Linda Wolfe, and Rebecca Zarger.

I am incredibly grateful to Michelle Bezanson, Lara Braff, Kerry Dore, Amanda Ellwanger, Katharine Jack, Susan Lappan, James Loudon, Katherine C. MacKinnnon, and Jeffrey Peterson who kindly read select chapters and provided insightful and constructive feedback that helped refine my thinking and greatly improved this manuscript. Michelle Bezanson deserves special recognition for creating the cover art that brings this book to life. I am also thankful to the anonymous reviewers who read my prospectus and manuscript, and to the editorial team (Katherine Ong and Marc Stratton) at Routledge.

Over the course of my career I have been very fortunate to work with so many outstanding students. Discussions in my ANTH 501, Primate Behavioral Ecology, and ANTH 601, Graduate Seminar in Biological Anthropology, courses at San Diego State University (SDSU) helped me cultivate a number of the arguments I make in this book. My field research has been enriched by the intellectual and logistical contributions of my graduate students, Amanda Ellwanger, KT Hanson, Kristen Morrow, Jeffrey Peterson, Jaima Smith, Tiffany Wade, and Alison Zak, each of whom were committed to the rigor and success of our projects, and who impressed me with their dedication to learning how to do fieldwork not just as primatologists, but as *anthropologists*. Many thanks go to the undergraduate interns in the "Primate Wing" at SDSU who assisted with data entry and analysis, and who conducted literature reviews that contributed to this book: Cristina Alcaraz, Kim Caliliw, Kira Collins, Kate Jameson, Noe Guerrero, Guillermo Guzman, Nancy Guzman, Morgan Miller, Meagan Platt, Bridget Rickman, Joshua Trinidad, and Devon Yuwiler.

My field research on the human-primate interface has been made possible with the logistical and financial support of a number of people and institutions.

Central Sulawesi, Indonesia: Noviar Andayani, Bambang Suryobroto, the Indonesian Institute of Sciences (LIPI), and Lore Lindu National Park sponsored my research; Duncan Neville, Shadiq, and David Lamanyuki provided logistical support. I am eternally grateful for the patience, enthusiasm, friendship, and hard work of my core team of field assistants, Manto, James, Papa Denis, Pias, Pak Asdi, and Tinus. South Sulawesi, Indonesia: many thanks go to my long-term research counterpart, Ngakan Putu Oka, as well Nenny Babo, Muhammed Restu, Isra Wahid, M. Iqbal Djawad, Andang Suryana Soma, Dwia Aries Tina Pulubuhu at Universitas Hasanuddin, Makassar, staff at Bantimurung-Bulusaraung National Park (Dedy Asriady, Iskandar Kamaruddin, Erista Murpratiwi, Adb Rajab), and fellow SulSel macaque researchers Monica Carosi, Cristina Sagnotti, Alessandro Albani, and Lavinia Germani for their support. Navigating the karst forests of South Sulawesi would not be possible without the assistance and good humor of my field assistants, Pak Haro, Hendra, Paisal, and Amir. I am also thankful for their friendship and long-term commitment to my field research and the livelihood of "our" macaques. Silver River, Florida: Alice Bard and Sally Lieb provided permission to conduct the research, Eben Kirksey spearheaded the project, and Bob Gottschalk, Mickey Summers, "Captain Tom" O'Lenick shared their knowledge and experiences navigating the Silver River. I am grateful to the National Science Foundation (NSF), Wenner-Gren Foundation for Anthropological Research, the Wildlife Conservation Society, the National Geographic Society, the American Society of Primatologists, the American Institute for Indonesian Studies (AIFIS), and the President's Leadership Fund and University Grants Program at San Diego State University for funding my field research. Critical Thinking Grants from SDSU's College of Arts and Letters provided course releases/protected time needed to complete portions of this book.

As this project was my first sole-authored book endeavor, I was admittedly unprepared for the level of focus and protected time I would need to see the project through to the end. Fortunately, I eventually figured out that I could be my most productive during personal writing retreats where I could be alone with my thoughts, my notes, my highlighters and favorite pens, and my tiny yet trusty Macbook Air®, with one requisite feature – mountains. Hence, I am grateful to southern California for providing the remarkable vistas – from the foothills of the Santa Ynez mountains in Santa Barbara, to Martha & Ted's "treehouse" in the San Bernardino mountains, to the nearby foothills of San Diego County – that gave me the peace, beauty, and inspiration to think and write.

Outside of academia, I am grateful for the support of my dear friends and family. My parents, Millie and Paul Riley, to whom I dedicate this book, cultivated my thirst for exploration and enabled me to learn early on how to trust in myself. They also provided unwavering support for each and every endeavor upon which I embarked. My sister, Rachele Riley, and "my persons," Nonie Kimpitak and Katie Leighty, are my rocks. My Beloit crew has always had, and will always have, my back. Kathy Peck made SDSU feel like home from day one; I treasure how

I always feel better when in her presence. My San Diego dance family kept my spirit high, my body strong, and my soul alive and well throughout the writing of this book. Finally, I thank my husband, Jason Ira Morgan, for sustaining our life together throughout this process, for always believing in me, and for being my biggest supporter. Thank you, *sayang*!

1

INTRODUCTION

The Promise of Contemporary Primatology

Primat-*ology*: the study of the members of the taxonomic Order, Primates. Primates are so special they have their own field of study. I get to call myself a primatologist, and (most) people know what this means. And, there are scholarly journals and societies whose explicit focus is primatology (e.g., the *International Journal of Primatology* and the International Primatological Society, respectively). But why are primates so special? Why do we have "primat*ology*" but not "Erinaceamorph*ology*" (i.e., the study of hedgehogs and their relatives)? The simple answer is because the order Primates includes us, humans. In the U.S., for example, primatology was given a home within anthropology, specifically in the subfield of biological anthropology, from its outset. This beginning dates back to the 1950s when physical anthropologist Earnest Hooton recognized the importance of primate studies for anthropology.[1] In his landmark paper in support of primate studies, Hooton (1954: 180) contended that the term "primatology" has a "…dual connotation: the study of the highest animal order and the study which of all the –ologies is, or ought to be, *primus*, prime, or first in importance."[2] His student, Sherwood Washburn (1973) later published a paper titled "The Promise of Primatology" in which he offered his vision of what primatology is and what he hoped it would become. Collectively, their vision was that the study of humans' closest living relatives, the nonhuman primates (hereafter primates), could shed light on human origins and the evolution of human behavior. And there you have it: a field devoted to the study of primates – their ecology, behavior, anatomy, physiology, and genetics across time and space – was born.

Compared to anthropology, primatology is a young field, and yet, so much has changed in terms of the context and conditions for the study of primates in the six or so decades since its birth. Borrowing from my academic great-grandfather Washburn, my goal in this book is to articulate what I see as "the promise of *contemporary* primatology."[3] With this goal, my focus is specifically on behavioral primatology in field settings. At the same time, this book is not meant to be an exhaustive coverage of all foci of today's behavioral primatology. Moreover, as a primatologist trained in the U.S., I note that my perspective primarily draws from the North American tradition of field primatology. While exploring the trends in contemporary primatology within different national traditions would be a fascinating endeavor, it is beyond the scope of this book. I do, however, weave in critical insights from other traditions when appropriate (e.g., Japanese schools of primatology).[4] I also acknowledge that much of contemporary primatology – from molecular primatology to evolutionary anatomy to sensory ecology, to name a few – continues to follow and build upon the fundamental vision laid out by Hooton and Washburn; indeed, the promise of today's primatology is multifold (see Higham & Dominy 2018). But, these important bodies of work are not my focus.

Building upon my article "Contemporary Primatology in Anthropology: Beyond the Epistemological Abyss" (Riley 2013), the promise of contemporary primatology that I argue for in this book is about forging an expanded purpose for primatology – one that situates it not just within the subfield of biological anthropology, but within the field of anthropology as a whole. The contemporary primatology that I highlight herein recognizes the reality of the anthropogenic context in which primates live and uses a broadened lens that includes the human dimension. This contemporary primatology tells us what it means to be a primate and live alongside other primates, nonhuman and human; it elucidates how humans and other primates behave together, co-shaping each other's ecology, sociality, and evolutionary trajectories; it tells us what primates (human and nonhuman) mean to each other. It also contributes to broader anthropological questions of what it means to be human and why humans do what they do. For example, why do people tolerate crop-feeding behavior more so from primates than other wildlife? To probe the human condition, we need to understand humans as they relate to and interact with other animals, including our closest living relatives.

My motivation for writing this book is multifold. First, despite the reality that primates are living in spaces increasingly modified by people, attention to the human dimension in existing general texts on primate behavior and ecology is largely relegated to concluding chapters on the plight of remaining primate populations.[5] Early on, Washburn himself recognized humans as primate predators and noted the considerable time depth of that role (Washburn & Hamburg 1968). Although in the last two decades growing attention is being paid to anthropogenic impacts on primates, humans are generally viewed as extrinsic factors, rather than as key components of the ecological communities in which primates live. The implications of this difference in perspective are significant, for the latter view allows us to more fully appreciate the role of humans in effecting evolutionary change. In this book, I provide examples of how and why actually studying the anthropogenic context is integral to our understanding of a whole suite of primate characteristics – from aspects of their biologies and ecologies to their conservation status. In doing so, my goal is to illustrate how our ability to investigate and interpret the adaptive significance of primate behavior is not compromised by the anthropogenic touch.

Second, even though the "science wars" are technically behind us, perceptions that a wide epistemological gulf exists between primatology and sociocultural anthropology, and that evolutionary approaches provide the sole means for a unified anthropology, persist.[6] With this book, I hope to dissolve these perceptions and motivate anthropological primatologists to continue exploring interesting points of intersection across anthropology in the twenty-first century and beyond. Biological anthropologist Karen Strier once remarked that "being a primatologist in an anthropology department sometimes felt like being the lone gorilla in zoo – no one remembered how we got there, there was no plan to provide us a companion, and they probably weren't going to replace us when we died" (2011: 7). The time has come for such sentiments to truly become a thing of the past.

This book is also motivated by the revolutionary vision outlined by Washburn (1951, 1953) with his "The New Physical Anthropology" and its more modern manifestation as "The New Biological Anthropology" (Fuentes 2010).[7] As part of the 15th Cold Spring Harbor Symposium on Quantitative Biology, "Origin and Evolution of Man," that Washburn co-advised in 1950 with evolutionary biologist Theodore Dobzhansky, Washburn articulated his vision for what he called "The New Physical Anthropology." Washburn (1951, 1953) argued for a shift in strategy away from mere description, measurement, and classification toward one focused on process and function and the use of evolutionary theory as an explanatory framework. Importantly, Washburn was not shunning description per se; rather, he argued that it should not be the end point. It should lead to an analytic stage focused on understanding the processes that led to the form, whether physical or behavioral, under study: "Emphasis on adaptation helps in the evaluation of the structures which are used in classification, and the constant emphasis on function and behavior frequently leads to the interpretation of descriptive data" (Washburn 1963: 203). He called for replacing the static, typological approach, which was 80 percent measurement with a more dynamic, theory-driven one that was 80 percent process. While these elements of Washburn's vision along with his emphasis on experimentation are widely cited, it is important to not lose sight of the fact that Washburn also recognized the importance of interdisciplinary work: to "understand the process of human evolution we need a modern dynamic biology and a deep appreciation of the history and functioning of culture" (Washburn 1953: 726). This appreciation of contingency was also apparent in his view of primate behavior:

> in the study of the nonhuman primates, we assume that behaviors are the result of the biology of the species, the ecology, the particular individual,

> and local history.....A gibbon's behavior is not at one level and its social life at another.
>
> *Washburn 1973: 178*

Therefore, Washburn envisioned a new physical anthropology that was simultaneously a biological science and a social science.

In the current era we see that his vision is alive and well, if not more aptly represented as "The New Biological Anthropology," which anthropologist Agustín Fuentes (2010: 5) considers a "more accurate, encompassing view of what we do." Today, we continue to build upon the vision of Washburn and other founders but, importantly, we are also pushing past old boundaries, creating expanded theoretical landscapes and new methodologies from linkages across anthropology's subfields and allied disciplines (e.g., biology, ecology, sociology, and science studies). My objective is for this book to reflect on this new, exciting path. It explores evolving paradigms and the blurring of borders, conceptual and disciplinary. It identifies points of intersection and spaces for collaborative exchange within the field of anthropology as well as beyond. This book is inherently reflexive; I am an anthropological primatologist looking in, turning my gaze toward the field that I myself practice, and in turn, hope to shape.

My Academic Coordinates

To appreciate my perspective and arguments, let me first clarify my academic coordinates, which lie at the intersection of primatology, environmental anthropology, and conservation. I'm a hybrid of sorts. I became interested in anthropology in high school when taking a class on "Classical Civilizations" that was offered as part of the International Baccalaureate (IB) program that I was fortunate to be enrolled in. It was because of this class that I became obsessed with arguably one of the most incredible archaeological discoveries: the life-sized terracotta soldiers that formed a wall of protection around the 1st Emperor of China, Qin Shi Huang Di's, tomb in Xi'an, China. Archaeology it was; well, until it wasn't. All it took was one day as a volunteer at an archaeological dig during a hot, sticky summer in southern Maryland for me to change my mind super quick. I remained nonetheless committed to pursuing an anthropology degree. So, in my first semester as an undergraduate at Beloit College, in lieu of taking the introduction to archaeology course, I enrolled in biological anthropology. Learning about fossil hominins captivated me, but, like archaeological material, hominin remains must also be "dug up."[8] Nope. Oh, but anthropologists also study living primates of the nonhuman kind? Sold.

Besides my newly found love for primates, I was also passionate about international travel and was particularly keen on studying abroad in a place outside of my comfort zone. It just so happened that one of the sociocultural anthropologists in the anthropology department at Beloit College was leading a study abroad to Zimbabwe. It didn't take much convincing; I was in. My professor and soon-to-be program leader knew I was interested in primates, and figured I might also be interested in conservation, so he suggested a book for me to read: *The Myth of Wild Africa* by conservation scholars Jonathan Adams and Thomas McShane (1992). This book heavily influenced my thinking about conflict and coexistence between people and wildlife and the practice of conservation, and so much so that I chose to focus my independent research project that semester on the emerging community-based conservation efforts in Zimbabwe rather than on its primates. A seed was laid.

Upon graduating from college, I was eager to get my feet wet in primatology, so I immediately accepted a position as a behavioral care technician at a large breeding facility for primates in South Carolina. It was here that I developed my love for macaques. The facility was macaque-tastic: it housed five species (*Macaca mulatta*, *Macaca fascicularis*, *Macaca nemestrina*, *Macaca maura*, and *Macaca sylvanus*) numbering in the thousands. I remember being completely overwhelmed when I first got there, thinking: how will I learn them all? It certainly took some time, but it was here that I not only learned a lot about macaque social behavior and the trials and tribulations of captive care and management, but also about the macaques as subjective beings. Conducting daily behavioral observations, managing the reintroduction of individuals into their social groups, and merging social groups over a period of three years meant that I came to know (almost) every single one of them as individuals. This appreciation of the subjectivities of our primate subjects continues to inform how I approach primate research.

After three years of captive work it was time for me to move on, so I began looking at graduate programs in primatology. When doing so, all of the literature I read and the experiences I gained during my study abroad program kept resurfacing in my mind. I realized that I didn't want to just study primate behavior; I wanted to study it in a broader context that included people and the environment. That context was provided by the Ph.D. program in Environmental and Ecological Anthropology at the University of Georgia (UGA). Through this program I took coursework in biological anthropology, sociocultural anthropology, historical ecology, ecological anthropology, primatology, conservation ecology, ecological economics, and tropical ecological and cultural systems. Hybridity. It was also during graduate school that I was introduced to the newly emerging approach of ethnoprimatology (Sponsel 1997; Wheatley 1999; Fuentes & Wolfe 2002) – the study of the human-primate interface – and my interest in conducting anthropological work that is integrative across the sub-fields, and has both fundamental and applied significance, became solidified.

For my dissertation research, I used the ethnoprimatological approach to examine the ecological and cultural interconnections between Tonkean macaques (*Macaca tonkeana*) and the human residents of the Lindu highland plateau in Lore Lindu National Park, Central Sulawesi, Indonesia (Riley 2005).

Over a 22-month period, I addressed questions concerning the impact of anthropogenic habitat alteration on the ecology of Tonkean macaques, how the macaques figured into local folklore, and the conservation implications of overlapping resource use between villagers and the macaques. This multifaceted project required an intra- and interdisciplinary methodological approach involving behavioral observations, analysis of habitat structure and composition, ecological surveys of cacao gardens for evidence of crop feeding by wildlife, and ethnographic interviews with farmers and other villagers. To accomplish all of this I was fortunate to receive funding from a number of sources.[9] I was also able to do so because of an amazing group of field assistants, "Team *Macaca*," who taught me what fieldwork – primatological and anthropological – is all about, and who continue to inspire me to think critically about many of the issues I raise in this book (Figure 1.1). Sad to leave behind my new home, I left Lindu with a heavy heart, but also with eagerness to share what I learned about the special place that is Lindu. Upon successful completion of my dissertation, I earned a graduate certificate in Conservation Ecology and Sustainable Development from the UGA's Institute of Ecology and a Ph.D. in Anthropology at UGA. In 2006, I became a faculty member in the Department of Anthropology at San Diego State University (SDSU), where I remain today.

FIGURE 1.1 "Team *Macaca*" Left to right: James, Papa Denis, Pias, Manto, and Erin. We are joined by Rifan, Papa Denis' son, who is holding a picture of "Mogly," the alpha male of one of our study groups of Tonkean macaques (*Macaca tonkeana*). Team members not present: Pak Asdin, Adam, Rahmat, Pak Bakti, Papa Tri), circa 2004.

Overview

The promise of contemporary primatology that I argue for herein is about recognizing humans as integral components in the ecologies of primates; using a broadened theoretical lens and methodological toolkit to study primate behavior and ecology in increasingly anthropogenic contexts; seeking points of intersection and spaces for collaborative exchange within the field of anthropology as well as beyond for both scholarly *and* applied purposes, such as primate conservation; and, forging an expanded purpose for primatology – one that situates it not just within the subfield of biological anthropology, but within the field of anthropology as a whole.

The first two chapters of the book that follow this introduction provide disciplinary and historical context for this promise. Chapter 2, "Franz Boas, American Anthropology, and the Biological-Sociocultural Divide," starts us off. In the U.S., primatology was given a home within the discipline of anthropology – a discipline that, largely due to the efforts of Franz Boas, encompassed a four-field approach to the study of the human condition: archaeology, biological anthropology, linguistic anthropology, and sociocultural anthropology. Despite this ideal structure, biological and sociocultural anthropology have had coexistence issues that run deep. With the goal of understanding how these threads of American anthropology came to be antagonists, I explore Boas' perspectives and efforts in the development of American anthropology. I then consider how the American Anthropological Association (AAA) and its flagship four-field journal, the *American Anthropologist*, have weathered the tension. I conclude this chapter by beginning to reflect on how the biological-sociocultural divide might be transcended. Namely, I argue that anthropology need not, and should not, be guided by one single paradigm; it should instead embrace theoretical and methodological pluralism. By doing so, we (anthropologists) are better positioned to value the distinctive forms of hybridity that contemporary primatology has to offer.

In Chapter 3, "Primatology: Becoming Anthropology," I explore the emergence of field primatology and how it came to be a part of the American tradition of anthropology. The key players in this selective review include the early movers and shakers, Robert Yerkes and Ray Carpenter, the early physical anthropologists who secured primatology's home in anthropology, and the founding mothers and fathers of anthropological field primatology. While anthropological primatology was originally centered on the study of primates as referential models for our early ancestors, it later expanded to encompass a broad comparative approach aimed at uncovering behavioral similarities and dissimilarities across the primate order and using these findings to better understand how evolutionary principles and processes have affected human biological evolution. I also explore how early anthropological primatology in the field was conducted, which in turn, exposes some of its methodological affinity with sociocultural anthropology. I conclude this chapter by arguing that the promise of contemporary primatology lies in its articulations across the field of anthropology and with recent trends in allied fields (e.g., evolutionary biology, conservation biology). As I show in subsequent chapters, these articulations are critical in order to comprehensively examine and explain primate behavior in the increasingly anthropogenic contexts in which extant primate populations currently live.

In the 1970s and 1980s, the field of primatology transformed as it sought theoretical and methodological guidance beyond the domain of anthropological practice, and primate behavioral ecology, an evolutionary approach derived from ethology and biology, came to dominate. Chapter 4, "Primatology in Anthropogenic Contexts: An Expanded Evolutionary Approach," explores this transformation. At the same time, primatologists were increasingly encountering primate populations that were not living in environments free from human influence. Does this reality constitute a fundamental problem for field primatology or does it provide opportunities for an expanded scope of study? I argue for the latter and highlight why it is important to study primate behavior in such settings: to explore primate adaptability at work! To do so, however, requires drawing from an extended evolutionary approach that adopts a dynamic view of adaptation, aims to explain behavioral variation in contemporary environments that include the human dimension, and incorporates additional principles and processes at play, such as behavioral plasticity and niche construction.

Recognizing anthropogenic contexts as worthy settings in which to examine primate behavior and ecology also characterizes the nascent field of ethnoprimatology, which is the subject of Chapter 5, "Primatology in Anthropogenic Contexts: The Emergence of Ethnoprimatology." Ethnoprimatology was the brain child of an ecological and sociocultural anthropologist in the late 1990s, but it has since become an important and increasingly used approach for field primatology (Dore et al. 2017). In this chapter, I review the development of the ethnoprimatological approach, highlighting its diverse scope of study that includes the ecological causes and consequences of human-primate sympatry, human-primate disease ecology, and cultural dimensions of the human-primate interface. I explore how research conducted using this approach can contribute to key concerns in biological anthropology, and how this approach enables boundary crossing. Both human and primate perspectives need to be understood as well as the points of intersections between them to comprehensively study the human-primate interface. To do so, we need theoretical and methodological pluralism. The ethnoprimatological approach, which integrates theory and technique from behavioral ecology, ecological and environmental anthropology, sociocultural anthropology, and conservation ecology, provides the pluralistic framework needed to study primates in the current era (Riley 2018).

Chapter 6, "Beyond the Divide: Fieldwork, Reflexivity, and Multispecies Worlds," builds on the theme of boundary crossing that permeates previous chapters. Here I explore a new frontier for anthropological primatology: intersecting with humanistic fields of study. I expose the epistemological affinities that primatology shares with sociocultural anthropology, such as the importance of being reflexive in our work, and what further insight primatologists can gain from reaching across the aisle. For example, I show how being cognizant of context and contingencies helps us better interpret the patterns of primate behavior we observe. Because it is becoming increasingly difficult to study a primate population that is not in some way affected by anthropogenic factors, primatology's domain of study is now more so than ever multispecies. We therefore need to examine how humans and other primates behave together, and how they interface and change one another, socially, cognitively, and biologically (Lestel et al. 2006). To effectively do so, we must engage with the perspectives and approaches emerging from the bourgeoning realm of "human-animal studies." This multidisciplinary and interdisciplinary body of work aims to uncover the complex and multifaceted ways human lives intersect with, influence, and are affected by non-human animals (Hurn 2017). When we draw from this work we are afforded, for example, the opportunity to recognize primate subjectivities and see them as agents, which in turn helps us to better interpret and explain their behavior. Just as human-animal studies scholars aim to understand *the human* as emergent through their relations with nonhuman beings (Ogden et al. 2013), contemporary anthropological primatology can draw from these approaches to understand *the primate* as emergent through its relations with other beings, including humans.

In Chapter 7, "Primate Conservation in the Twenty-First Century and Beyond," I address the remaining dimension of the promise of contemporary primatology as I see it: actively engaging in the conservation of primates, and doing so in an integrative way that draws from the natural sciences, the social sciences, and the humanities. In recent years, field primatologists are increasingly committed to aiding efforts to preserve remaining primate populations. I would argue that in the current era it would be unethical not to, but we need guidance. In this chapter, I explore what we primatologists can learn from our fellow scholars, from the natural sciences *and* the social sciences and humanities, on how to move toward preventing the irreparable loss of our closest living relatives (and our study subjects). As anthropogenic activities continue to alter the environments in which primates live, understanding how they are able to behaviorally respond to these impacts is a critical component for designing conservation management efforts. At the same time, conservation is not just about forests, trees, or primates; it is also about people. Primatologists can therefore draw from broader social science and environmental humanities scholarship on the practice of conservation. With Chapter 8, I conclude the book by returning to the promise Washburn outlined 45 years ago. Namely, I explore how the promise of contemporary primatology I argue for herein compares to Washburn's. I also consider whether Washburn would be pleased with what primatology is today, and what it will become.

This book supports a specific promise for contemporary anthropological primatology – one that accepts the increasingly anthropogenic context in which primates live as a worthwhile setting in which to explore primate adaptability at work, and one that relies on a cross-field and cross-disciplinary exchange and a pluralistic theoretical and methodological grounding. While the "we" I use in this book is primarily meant as "we primatologists," I simultaneously consider it an inclusive we: we who care about primates, about people, and about a future in which humans and other primates can sustainably coexist.[10] Accordingly, this book is meant for primatologists and biological anthropologists, but my hope is that students and scholars interested in integrative anthropological research, sociocultural anthropology, environmental anthropology, the history of science, sciences studies, and conservation and evolutionary biology will also find it useful.

Notes

1 A note about terminology. My personal preference is to use "biological" rather than "physical" to describe the field and its scholars because the former better captures the scope of the field and dynamic nature of its foci. However, I use "physical" when referring to individuals who either practiced at a time when it was "physical anthropology" and/or whose work encapsulates what "physical anthropology" connotes (i.e., an emphasis on physical traits, types, and measurement).

2 Following Hooton (1954), biological anthropologist Karen Strier (2011) notes that if the structuring of academic disciplines were to follow the Linnean classification system, then anthropology would actually be a subfield of primatology, not vice versa. I have to admit; I like this idea. Hooton (1954) himself actually called for the establishment of "Departments of Primatology," which of course never came to fruition in the U.S. (or at least not yet). Interestingly, Linneaus' first classification of primates actually followed the departmental structure of anthropology with primatology as a subfield: in his first edition (1735) of *Systema Naturae*, Linneaus classified all primates under *Anthromorpha*, meaning "man form." By 1758 though he changed the name of the order to the one that holds today, Primates.

3 Washburn is my academic great-grandfather through two paths: Washburn's students, Phyllis Dolhinow (Berkeley) and Claud Bramblett (UT Austin) were, respectively, the PhD advisors of my undergraduate mentor (Nancy Krusko) and my PhD advisor (Carolyn Ehardt). See also Figure 3.3.

4 See Ribnick (1982), Asquith (1991, 1996), and Takasaki (2000) for review of the primate field research traditions in Japan.

5 A notable and recent exception is Karen Strier's (2017) 5th edition of *Primate Behavioral Ecology* in which she covers ethnoprimatology and the study of the human-primate interface in the chapter on community ecology. I should note though that there are a number of edited volumes that explicitly address anthropogenic impacts on primates and the human-primate interface. Additionally, although not technically a general primatology textbook, Strier's (2014) *Primate Ethnographies*, a collection of "autoethnograhic" essays written by over a dozen established field primatologists, purposefully highlights the human dimension of primate fieldwork.

6 Another note about terminology. I prefer to use the term "sociocultural anthropology" over "cultural anthropology" because the former is more comprehensive, explicitly combining the multiple traditions (American, British, French) that encompass the field. I therefore use "sociocultural anthropology" throughout the text unless I am specifically referencing the work of scholars who refer to it as "cultural anthropology."

7 It is worth noting that Washburn's chapter in J.O. Brew's 1968 edited volume, "One Hundred Years of Anthropology," was titled "100 years of *Biological* Anthropology" (my emphasis). It's not clear though who was responsible for the titling: Brew or Washburn, or Harvard University Press, for that matter.

8 Fossil hominins, in particular the "Black Skull" (then *Australopithecus*, now *Paranthropus aethiopicus*), enthralled me so much that I got a tattoo of it on my right ankle at age 19. It's a terrible tattoo, but it is rich in meaning, so it will remain…for life. Dr. Alan Walker, who discovered the Black Skull, died the day I wrote this footnote. Rest in peace, Dr. Walker; your tremendous impact on anthropology will carry on.

9 My research was funded by a Doctoral Dissertation Research Improvement Grant (DDRIG) from the National Science Foundation, the Wenner-Gren Foundation for Anthropological Research, the Wildlife Conservation Society, an American Society of Primatologists Conservation Grant, and the Melissa Hague Foundation.

10 I love that the Indonesian language (Bahasa Indonesia) distinguishes the inclusive "we": *"kita" bukan "kami."*

References

Adams JS, McShane TO. 1992. *The Myth of Wild Africa: Conservation Without Illusion.* University of California Press.

Asquith PJ. 1991. Primate research groups in Japan: Orientations and East-West differences. In *The Monkeys of Arashiyama: Thirty-five Years of Research in Japan and the West*, ed. L Fedigan, PJ Asquith, pp. 81–98. New York: SUNY Press.

Asquith PJ. 1996. Japanese science and Western hegemonies: Primatology and the limits to set questions. In *Naked Science: Anthropological Inquiry into Boundaries, Power, and Knowledge*, ed. L Nader, pp. 239–56. New York: Routledge.

Dore KM, Riley EP, Fuentes A. 2017. *Ethnoprimatology: A Practical Guide to Research at the Human–Nonhuman Primate Interface.* Cambridge: Cambridge University Press.

Fuentes A. 2010. The new biological anthropology: Bringing Washburn's'New Physical Anthropology into 2010 and beyond--The 2008 AAPA luncheon lecture. *Yearbook of Physical Anthropology* 53: 2–12.

Fuentes A, Wolfe LD, eds. 2002. *Primates Face to Face: The Conservation Implications of Human-Nonhuman Primate Interconnections.* Cambridge: Cambridge University Press.

Higham JP, Dominy NJ. 2018. The promise of primatology fulfilled? *American Journal of Physical Anthropology* 166: 783–90.

Hooton E. 1954. The importance of primate studies in anthropology. *Human Biology* 26: 179–88.

Hurn S. 2017. Human-animal relations (HAR). In *International Encyclopedia of Primatology*, pp. 1–7: John Wiley & Sons, Inc.

Lestel D, Brunois F, Gaunet F. 2006. Etho-ethnology and ethno-ethology. *Social Science Information* 45: 155–77.

Linnaeus CV. 1735. Systema Naturae, sive Regna Tria Naturae systematice proposita per classes, ordines, genera, & species. *Lugduni Batavorum: de Groot. Staffan Müller-Wille.*

Ogden LA, Hall B, Tanita K. 2013. Animals, plants, people, and things: A review of multispecies ethnography. *Environment and Society* 4: 5–24.

Ribnick R. 1982. A short history of primate field studies: Old World monkeys and apes. In *A History of American Physical Anthropology 1930–1980*, ed. F Spencer, pp. 49–73. New York: Academic Press.

Riley EP. 2005. *Ethnoprimatology of Macaca tonkeana: The Interface of Primate Ecology, Human Ecology, and Conservation in Lore Lindu National Park, Sulawesi, Indonesia*. Ph.D. Dissertation thesis. University of Georgia.

Riley EP. 2013. Contemporary primatology in anthropology: Beyond the epistemological abyss. *American Anthropologist* 115: 411–22.

Riley EP. 2018. The maturation of ethnoprimatology: Theoretical and methodological pluralism. *International Journal of Primatology* 39: 705–29.

Sponsel LE. 1997. The human niche in Amazonia: Explorations in ethnoprimatology. In *New World Primates: Ecology, Evolution, and Behavior*, ed. WG Kinnzey, pp. 143–65. New York: Aldine Gruyter.

Strier KB. 2011. Why anthropology needs primatology. *General Anthropology* 18: 5–8.

Strier KB, ed. 2014. *Primate Ethnographies*. Boston: Pearson.

Strier KB. 2017. *Primate Behavioral Ecology*. 5th Edition. London: Routledge.

Strum SC, Fedigan L. 1999. Theory, method, gender, and culture: What changed our views of primate society? In *The New Physical Anthropology: Science, Humanism, and Critical Reflection*, ed. SC Strum, DG Lindburg, D Hamburg, pp. 67–105. Upper Saddle River, NJ: Prentice Hall.

Takasaki H. 2000. Traditions of the Kyoto School of field primatology in Japan. In *Primate Encounters: Models of Science, Gender, and Society*, ed. SC Strum, L Fedigan, pp. 151–64. Chicago: Chicago University Press.

Washburn SL. 1951. The new physical anthropology. *Transactions of the New York Academy of Sciences* 13: 298–304.

Washburn SL. 1953. The strategy of physical anthropology. In *Anthropology Today: An Encyclopedia Inventory*, ed. AL Kroeber, pp. 714–27. Chicago: University of Chicago Press.

Washburn SL. 1963. Behavior and human evolution. In *Classification and Human Evolution*, ed. SL Washburn, pp. 190–203. Chicago: Aldine Publishing Co.

Washburn SL. 1968. One hundred years of biological anthropology. In *One Hundred Years of Anthropology*, ed. JO Brew, pp. 97–115. Cambridge, MA: Harvard University Press.

Washburn SL. 1973. The promise of primatology. *American Journal of Physical Anthropology* 38: 177–82.

Washburn SL, Hamburg DA. 1968. Aggressive behavior in Old World monkeys and apes. In *Primates: Studies in Adaptation and Variability*, pp. 458–78. Holt, Rinehart, and Winston: New York.

Wheatley B. 1999. *The Sacred Monkeys of Bali*. Prospect Heights, IL: Waveland Press, Inc.

2

FRANZ BOAS, AMERICAN ANTHROPOLOGY, AND THE BIOLOGICAL-SOCIOCULTURAL DIVIDE

My objective in this chapter is to do some contextualizing, some stage setting, if you will. To understand what contemporary primatology can be first requires an understanding of the disciplinary context in which primatology first emerged. In the U.S., that disciplinary context is American anthropology – a distinctive tradition that examines the human condition across four quadrants: archaeology, biological anthropology, linguistic anthropology, and sociocultural anthropology (Figure 2.1). While some scholars see American anthropology's four-field structure as a "historical accident" related to how it was institutionalized (Silverman 2005), others see considerable purpose in its formulation (Yanagisako 2005). American anthropology was originally focused on the study of the native peoples of North America. Accordingly, a four-field approach was a necessary methodology to employ in order to conduct a comprehensive study of Native Americans; that is, their physical forms, archaeological history, language, and customs (Bourguignon 1996). American anthropology of today, of course, has a much more diverse scope, but over the years, the nature of its distinctiveness – the four-field approach or the more pejorative "sacred bundle"[1] – has not been without controversy and contention. The origin and essence of the American tradition of anthropology has been the subject of considerable reflection and scholarship,[2] so my objective here is a bit more modest. I aim to look specifically at the nature of the relationship between the biological and sociocultural threads of the tradition. Even still, this is a sizable endeavor. I therefore focus primarily on the perspective and efforts of the anthropologist who championed the four-field approach and who is widely recognized as the "Father of American Anthropology," Franz Boas. A significant body of work has examined and highlighted Boas' contributions to anthropology, so my aim here is not so much to retell those contributions. Rather, I focus on Boas as a way to retrospectively examine how the divide between biological and

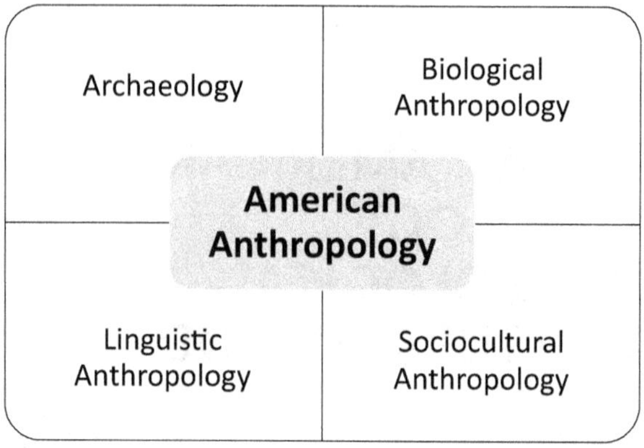

FIGURE 2.1 The four quadrants of American anthropology.

sociocultural anthropology, largely fueled by an underlying rift between science and humanism,[3] came to be. Anthropology is, after all, "the most scientific of the humanities and the most humanistic of the sciences" (Wolf 1964: 88). This analysis will in turn provide fodder for further contemplation of contemporary anthropological and primatological practice in that chapters that follow. In other words, to figure out how to move beyond the divide, we need to know a little bit about it.

Franz Boas: You Can Call Him "Papa"

Although Franz Boas (1858–1942) is largely remembered as a sociocultural anthropologist, his academic coordinates originated in physics, and many historians of science will remind us of the important role he played in the development of physical (biological) anthropology in the U.S. (e.g., Darnell 1982; Little 2010). In particular, Boas' contributions to studies in human growth and development were many: he developed more sophisticated statistical procedures to examine anthropometric data (e.g., Boas 1894), and as highlighted by Darnell (1982) and Little (2010), his work showing that the tempo of growth varies in relation to sex and age expanded our understanding of the complexity of human development (e.g., Boas 1932, 1935). Another substantive, and in many ways, game changing, contribution was his work exploring the effect of the environment on human morphology. His now famous research (Boas 1912) documenting changes in stature and head shape in American-born immigrants compared to their European-born parents and exploring the powerful role of demographic and socioeconomic factors shaping human biology clearly demonstrated the plasticity of physical traits thought to be markers of human "races," thereby calling into question the stability of the static notion of biological race.[4]

In retrospect Boas was therefore ahead of his time, foreshadowing through his specific perspectives and research foci, the future trajectory of biological anthropology. Take, for example, the name of the subfield. Anthropologists, if anybody, know quite well that words matter, as they say a lot about what people find important. Biological anthropology, for example, is technically the same field as physical anthropology, but the current preference[5] for "biological anthropology" reflects our commitment to viewing the subfield as focused on the dynamic processes of human biological variation and evolution rather than a static view of human variation imbued by "physical anthropology." Interestingly, in the early 1900s Boas was referring to the subfield as "biological or somatic anthropology" (1904: 518) only later adopting "physical anthropology" as his terminology of choice.[6] More substantively, in his view on the study of the distribution of human types, Boas foreshadowed Washburn's call for a move away from mere measurement and static typology toward an examination of process: "No matter what the ultimate explanation of the distribution of types may be, we cannot evade the task of investigating their present distribution and of seeking for the explanation for the *reasons* for such distribution" (Boas 1899: 168; my emphasis added). And, while we're at it, let's certainly not forget that through his work on human growth, development, and the plasticity of the human form, Boas was essentially arguing for what we now call a "biocultural approach." Boas was, as aptly stated by biological anthropologist Michael Little (2010: 78), a "truly modern anthropologist scientist who arose during the 19th century."

Despite these important contributions to biological anthropology, perceptions regarding Boas' overall impact on the field vary. For example, historian of science George Stocking Jr. (1968: 188) describes Boas' immediate impact on the subfield as "limited," largely because he produced very few PhD students in biological anthropology and he did not participate in the key areas, such as genetics and inheritance, that were coming to define "the new physical anthropology" in the early to mid-1900s (Stocking 1974; Spencer 1981; Darnell 1982). Regardless, there is little contention over his impact on the broader field of anthropology. Boas is widely credited as the father of the four-field tradition of American anthropology.[7] This tradition is distinct because it contrasts greatly with the British and French schools of anthropology, which focus solely on social anthropology, with biological anthropology and archaeology deemed separate disciplines, and linguistic anthropology simply not existing.[8]

For those who train in the U.S. it may be easy to sometimes lose sight of this difference, but it is significant for it has played a major role in shaping the way that scholars think about and do anthropology. For those of us who conduct research outside of the U.S. context though, we are frequently reminded of discernible differences among the nationalist traditions, particularly as they have spread around the world. During my own fieldwork in Indonesia, for example, I am continually struck by people's reactions when I tell them that one, I am anthropologist, and two, I study nonhuman primate behavior and ecology. More often than not I am met with a look of immediate confusion because in Indonesia anthropology is about human culture, deriving from early interest by the Dutch in language, literature, and cultural history, or *Indologie* (Koentjaraningrat 1987). Archaeology is in its own department, biological anthropology, following the European traditions, is housed in anatomy departments and primatologists occupy positions in biology or forestry departments. But, I digress; back to papa Franz.

Boas' conception of the four-field approach – an "embracive anthropology" (Stocking 1974: 14) – has a complex origin, undoubtedly the result of the interaction of a diverse array of institutional, historical, and logistical factors (Stocking 1974; Darnell 1982, 2001; Silverman 2005). Upon returning to the U.S. after another year of study in Germany (1885–1886), Boas brought with him elements of the German anthropological tradition that included both biological and sociocultural anthropology from his teachers Rudolf Virchow and Adolf Bastian, respectively (Stocking 1974).[9] Boas subsequently found such an approach methodologically useful in his own research on the peoples of Native America. For example, in the fieldwork he conducted in Canada's Pacific Northwest for the British Association (1888–1897), Boas studied the languages and dialects, customs, social organization, and physical characteristics via anthropometric measurements of the Kwakiutl (Boas 1974 [1898]).

An embracive anthropology was also made possible by anthropology's institutional landscape at that time, including the founding of the Bureau of American Ethnology in 1879 by John Wesley Powell which organized and supported ethnographic, linguistic, and later, archaeological research on native peoples of North America, Harvard University's Peabody Museum of Archaeology and Ethnology (founded in 1866), and the American Museum of Natural History (founded in 1869), which began amassing and housing important anthropological collections (Darnell 2001). Finally, I would be remiss to not acknowledge, though not necessarily cited by Boas, that other late nineteenth-century scholars were also thinking in an inclusive way about the discipline of anthropology. That's right, we actually have many papas! In fact, some were much more explicit in their writings than Boas about their definition of anthropology and its scope. For example, Daniel Brinton (1892), a professor of archaeology and linguistics at the University of Pennsylvania, was concerned about the multiple ways the term "anthropology" was being employed and hence explicitly outlined his scheme "the nomenclature and classification of the anthropological sciences" (p. 265) as: I. Somatology (or Physical anthropology, II. Ethnology, III. Ethnography, and IV. Archaeology at an address to the Anthropological Society of Washington in 1892.[10] At the same address, Powell made the following remarks after Brinton concluded: "… in late years Anthropology has been used as a term to designate the whole science of man – man as an animal, man as a thinking bring, and man as an actor in economic arts, institutions, languages, literature, fine arts, religions and opinions. It is in this broad sense that I use the term Anthropology" (p. 268). Nonetheless, it is Boas who receives the most credit and whom we remember the most, which is likely attributed to his untiring dedication to the discipline and the "multifaceted research competence" he imbued (Stocking 1974: 14).[11]

Before the turn of the twentieth century, Boas was already talking about the different branches of anthropology, yet in his 1899 piece he only mentions three, namely biological, cultural, and linguistic – "the three branches of anthropology must proceed each according to its own method; but all equally contribute to the solution of the problem of the early history of mankind" (p. 171). His three major publications in 1911 – a) "Handbook of American Indian Languages," b) "Changes in the Bodily Form of Descendants of Immigrants," and c) "The Mind of the Primitive" – which Stocking (1974) sees as Boas' most important works – and his "Race, Language, and Culture" first published in 1940 also reflect where Boas' primary interests laid. That said, although not as extensively, Boas also contributed to the field of archaeology arguing for the importance of doing it alongside of ethnology and linguistic study (Boas 1902a). And, to be fair, by 1904 he defines anthropology as "[It is] the biological history of mankind in all if its varieties; linguistics applied to people without written languages; the ethnology of people with historic record, and prehistorical archaeology" (p. 523). What is often not necessarily discussed in standard accounts of Boas' four-field approach, however, is the fact the Boas himself seemed to waver on the stability of this tradition (but see Darnell 1982). For example, in that same 1904 address on the "History of Anthropology" given to the International Congress of Arts and Sciences, he summarizes the key problems of the branches of anthropology, argues that anthropologists doing fieldwork must be familiar with the principles of the each field's key methods "since all of them are needed for the investigation of his problems" (p. 524), and yet hints at the possibility of fission: "The time is rapidly drawing near when the biological branch of anthropology will be finally separate from the rest and become a part of biology" (p. 523). Boas' writings demonstrate that his wavering was largely concerned with the relationship between biological and sociocultural anthropology (e.g., Boas 1936). Boas (1936: 172) saw the drifting apart of these two subfields as "unavoidable" due to differences in research foci and methods. What is notable about his concerns is that he seemed to be primarily speaking to biological anthropologists[12]; he focuses on how biological/physical data can only be understood when considering ethnic background, environmental influences, and social conditions.

It is clear then that the tension between the biological and sociocultural subfields runs deep. As recognized by Boas, the source of the tension between these two quadrants of anthropological inquiry largely derives from the existence of two competing traditions that have informed how to do anthropology since the field's inception: objective and humanistic approaches (Rappaport 1994). Inspired by the biological sciences, the objective approach aims to discover causes or laws to explain phenomena, typically using the scientific method. While some threads of sociocultural anthropology utilize this approach (e.g., cultural materialism and human behavioral ecology), it is more biological anthropology's cup of tea. On the other hand, drawing from philosophy and the humanities, the humanistic approach seeks subjective knowledge and meaning-making through interpretation. Deep concern, dare I even say, fear, of the divide and the potential ramifications of it for anthropology that emerged at the outset of the disciplining of anthropology have persisted over the years. Anthropologists have invoked similes and metaphors of dismantling nation states, infrastructural tragedies, mental institutions, and flailing animals to describe the challenges that anthropology faces.[13]

One might ask though – should we really be all that surprised? – after all, even from outset of the American tradition, Boas himself foresaw potential fissures. Some have even questioned whether the divide is real, and whether hints at the field's fragmentation are exaggerated (Darnell 2002). These critiques center around claim that even beyond Boas' waverings, holism, that is, true integration of and collaboration across the four anthropology quadrants barely characterized the early years of American anthropology (Borofsky 2002).[14] If this is the case, what then exists to fragment and divide? Regardless, the ideal of four-field holism persisted, and as long as it continues to persist, tension and attempts at division are real and possible. Understanding the factors that triggered a resurfacing of the tension in more recent times (i.e., latter part of the twentieth century) and how the American Anthropological Association (AAA) and its journal, the *American Anthropologist*, weathered the tension is not only an interesting lesson in the history of anthropology, but also is needed to contemplate future transcendence of the divide, putative or real.

The AAA, the AA, and the Biological-Sociocultural Divide

In her 1993 *Science* piece titled "Anthropology: Nature-Culture Battleground," science writer Virgina Morell describes the most recent resurfacing of the biological-cultural divide occurring in the late 1970s following the emergence of E.O. Wilson's (1975) "Sociobiology: The New Synthesis" and the hinting that this adaptive framework that sees behavior as the result of an individual's drive to pass on their genes to the next generation could be applied to the human species. This assertion ruffled many anthropological feathers; and still does, for that matter. If all human behavior could be explained as the result of gene-maximization and selection, then what would this mean for the "stuff" of sociocultural anthropology – i.e., history, traditions, mythology, etc. – and its emphasis on behavior emerging in the context of social learning?

Around the same time, a more critical approach to anthropological inquiry began to take root in sociocultural anthropology.[15] Sociocultural anthropology was experiencing a "postmodern turn," whereby its fundamental core, ethnography, was critiqued as a tool that advanced hidden cultural values and political agendas rather than generating objective knowledge about why people do what they do. Sociocultural anthropologists' responsibility was therefore "to critique hidden and open oppressions of Western bourgeois culture: its racism, sexism, nationalism, homophobia, and scientism" (D'Andrade 2000: 223). To do so requires self-reflexivity and actively recognizing that ethnographic accounts are partial and incomplete, enact power relations, and imbue positionality (Clifford

1986). This perspective was firmly rooted by the mid-1980s. Meanwhile, biological anthropologists were being lured closer to the biological sciences by the latest tools in molecular biology (Morrell 1993: 1798). Further exacerbated by resource shortages within academic institutions, by the late 1980s to early 1990s, mutual intolerance was heightened, resulting in the realization of the divide at some academic institutions. At Duke University, biological and cultural anthropology split into two separate departments in 1988. In the early 1990s, some of the biological anthropologists in the Department of Anthropology at the University of California, Berkeley, moved to the Department of Integrative Biology. At Harvard University, following the creation of a new life sciences division in 2003, the biological anthropology faculty proposed splitting from the anthropology department to form a new department of human evolutionary biology within the new division, citing their desire to intellectually position biological anthropology closer to ecology, evolution, and organismal biology than the other three subfields (Shenk 2006). In 2009, that split was realized.[16] Stanford University's four-field anthropology department also suffered internal strife, leading to the creation of two separate departments in 1998: a sociocultural anthropology department for social and cultural anthropology led by those who have been vocal critiques of holism and the "sacred bundle" (e.g., Segal & Yanagisako 2005), and a four-field anthropological sciences department for the empirically-oriented (Shenk 2006). In 2007, however, the dean of humanities and science unilaterally reinstated a one-department model, much to the chagrin of some of the faculty (Jaschik 2007). Reunited, at least structurally.[17]

So how have the American Anthropological Association (AAA) and the *American Anthropologist* (AA) handled the divisiveness? As part of Boas' vision for the professionalization of American anthropology, the AAA was founded in 1902 (Stocking 1960).[18] Although the AAA was originally meant to include all of the subfields of anthropology, during the twentieth century sociocultural anthropologists gradually began to dominate the membership (Darnell & Gleach 2002). It was also during this time that the AAA experienced some restructuring. In 1946 a small group of individuals with executive power was formed along with a division of the membership into two types: qualified anthropologists ("members") and students and interested laypeople ("fellows") (Darnell & Gleach 2002: xviii). Further restructuring occurred in 1980s when a tax audit by the Internal Revenue Service compelled the AAA to merge (some of) the separate "quadrant societies" and other pre-existing units (now called sections) to become an umbrella organization (Stocking 1988). While on the one hand a sense of unity was fostered by this reorganization, it also precipitated increasing segmentation and specialization, which could simultaneously promote diversity with the field but also threaten the disciplinary unity the AAA was meant to uphold.[19] The stability of disciplinary unity was also ephemerally rattled in 2010 when the Executive Board of the AAA removed (but then later reinstated) the word "science" from various places in the mission statement of the AAA's long-range plan.[20]

Like the AAA, the *American Anthropologist*, the flagship journal of American anthropology, was also tasked with the responsibility of representing disciplinary unity. Cultural anthropologists Daniel Segal and Sylvia Yanagisako (2005) see this being particularly the case following the exit of many biological anthropologists upon the 1983 reorganization of the AAA.[21] In the face of increasing sectionalism and specialization, the idea was for the AA to be the place to illustrate the value of subfield integration and collaboration. Interesting ideas and questions whose spatial coordinates did not map onto just one the field's four quadrants were given a new space; a space to highlight intersections in a mutually intelligible way. In reviewing papers published in the AA from 1971–1995, anthropologist Regna Darnell (2002: 4) notes how struck she was with a trend toward greater subfield integration; one that she hadn't necessarily perceived at the time of the publications:

> As I became reacquainted with it I quickly discovered that the journal was not like I remembered it. I read most articles as they appeared but recall feeling no conviction that this single journal would, or could, or even should reflect the complexity of the discipline. The inclusion of diverse specialization and subdisciplines outstripped my recollections. I was gratified to observe that many articles defied subdisciplinary categorization.

While a review of the trends of the AA since that time is the beyond the scope of this text, a perusal of the editorials penned by the incoming Editor-in-Chiefs of the AA is illuminating. Here I build on the analysis presented by biological anthropologist, James Calcagno. As part of a 2003 special issue of the AA devoted to biological anthropology, in his piece titled *Keeping Biological Anthropology in Anthropology, And Anthropology in Biology*, Calcagno (2003) explores (via letters from the incoming editors) how the *American Anthropologist* weathered the increasing divide between biological and sociocultural approaches in anthropology during the "science wars" of the 1990s. As Calcagno notes, Janet Dixon Keller's (1990) editorial acknowledges the centralizing function of the journal and welcomes contributions from all subfields from both basic and applied domains.[22] In 1994, two sociocultural anthropologists, Barbara and Dennis Tedlock, took over as a two-person editor-in-chief team. These editors explicitly acknowledged the tensions that anthropology was facing at the time, and called for a ceasefire, so to speak. Their stated goal was to "keep this journal in the mainstream of anthropological thinking and research while simultaneously opening its pages to the many new voices and new approaches to the study of humankind" (1994: 521). While the word "unity" does not appear in their editorial, they do encourage authors to submit papers that would "tempt readers in neighboring fields to cross over into their own" (p. 521). Robert Sussman, a biological anthropologist (specifically, primatologist) who took over in 1998, uses similar "welcoming of all subfields of anthropological inquiry" language and made explicit his perspective that both qualitative and quantitative approaches have value and that "the combination of these approaches is what makes the results of anthropological research unique" (p. 606). He also reinstituted the Contemporary Issues Forum – an AA feature initiated in 1992 that aimed to "encourage productive discussion of current issues of relevance to anthropologists and the public" (Hill 1992: 809), but also one that Sussman (1998: 605) saw as a space for "bridging gaps between subdisciplines."

The editorials of the 2000s continue to encourage diverse perspectives and a balance across the subfields and across the basic and applied domains. However, a few features stand out to me in particular. The first is the absence of any language *explicitly* addressing tension or a divide within anthropology after sociocultural anthropologists Fran Mascia-Lees and Susan Lees' (2002: 713) editorial that states a commitment to publishing papers that "offer diverse perspectives cutting across the *polarities*" (my emphasis). Given that what we don't say is often just as important, if not more, than what we do say, does this absence suggest a waning divide? The second feature that stands out is that two of the editors during this decade, Benjamin Blount (2006), an environmental and cognitive anthropologist and Tom Bellstorff (2008), a sociocultural and linguistic anthropologist, explicitly state that in addition to upholding the unity mission of the journal, they also welcome submissions that *don't* necessarily adhere to the four-field vision. Both emphasize, however, the importance of nonetheless making these submissions intelligible, and hence, accessible to many. This too suggests that anthropologists were willing to move beyond the divide, particularly given that 14 years prior, the Tedlocks were heavily criticized as not being sufficiently loyal to the four-field model (Segal & Yanagisako 2005: 7). At the same time, the third noticeable feature is the explicit concern about the underrepresentation of biological anthropology in the pages of the AA, and an appeal to biological anthropologists to submit, submit, submit (Mascia-Lees & Lees 2002). This concern over the paucity of submission from biological anthropologists is echoed by Michael Chibnik (2013 a and b), who in addition to his "From the Editor" piece upon taking over in 2013, devoted a separate editorial to "Biological Anthropology in *American Anthropologist*." Mascia-Lees (2006) notes that biological anthropology represented only 3–5 percent of the total submissions she received between 1998–2003. From 2008–2016, of the 1091 manuscripts submitted to the AA, 88 (or 8% of total submissions) represented biological anthropology (Michael Chibnik, personal communication).[23] Of those 88 submitted, 13 were accepted for publication, and seven were never resubmitted (Michael Chibnik, personal communication).

Given that the AA is one of few generalist publications where anthropologists of any flavor can present their work, not to mention the only journal published by the AAA that publishes papers in biological anthropology and archaeology, it is curious why there have been so few submissions from biological anthropologists. This discrepancy could be a function of the fact there are simply more sociocultural anthropologists being awarded PhDs in the U.S. compared to the other subfields (Darnell 2002; Balée 2016). More bodies and minds in one area, means more potential submissions. Biological anthropology membership in the AAA has also been low; Mascia-Lees (2006: 13) notes that the percentage has never surpassed 5.35 percent.[24] One way to interpret these numbers is that biological anthropologists are less likely to be committed to a four-field model for anthropology (see Mascia-Lees 2006). Another likely explanation, particularly regarding the rates of submission to the AA, articulated by biological anthropologist Julienne Rutherford (2013), concerns a more practical challenge: given the diverse bodies of scholarship that biological anthropologists want to (and need to) reach, including scholars from the social and biomedical sciences, they may be more inclined to submit their best work to journals that are automatically indexed in databases (e.g., PubMed) that will ensure exposure and accessibility to said scholars.[25]

This review of editorials demonstrates that the editors of the AA, all of whom save one are sociocultural anthropologists, have continued to uphold a commitment to the four-field model over the last few decades. But what about beyond the pages of the AA? What perspectives emanate from the broader community of anthropologists? When sociocultural anthropologist Yolanda Moses (1997), then president of the AAA, polled the membership, 90 percent of respondents indicated a commitment to the four-field model.[26] Since that time, it seems that tensions have waxed and waned. Discussions of the divide have continued to pop up through more formalized means, such as the plenary session at the 2003 American Association of Physical Anthropologists meeting titled "Can biological and cultural anthropology coexist?" and a series of ensuing commentaries published in the January 2006 issue of *Anthropology News* and again in January 2010 (see Bruner 2010). And, let's not discount the informal discussions that likely occur over coffee breaks at annual conferences, in the hallways and behind closed office doors within anthropology departments, and via email and other forms of social media. Accepting the model passively and unreflexively (see Darnell 2009), if this was ever the case, appears to be a thing of the past. "Can we coexist?" is a perennial question that we just can't seem to stop asking, and, it's likely to continue to be asked as long as the four-field model remains in place. Indeed, among biological anthropologists there are some who would be delighted by a permanent fracture of our field so that they no longer have to "escape to biology departments."[27] Others, however, argue that without the cultural component biological anthropologists are little more than "egocentric zoologists" (Ruff 2000: 2). Sociocultural anthropologists also vary in their viewpoints toward preserving the sacred bundle.[28] While some argue that anthropology needs both scientific and interpretive approaches, and by implication, both biological and sociocultural quadrants (see Bourguignon 1996; Nader 2001; Ingold 2001), others question what the "biocultural synthesis" offers sociocultural anthropology itself, and have welcomed disciplinary and departmental divisions (e.g., Yanagisako & Segal 2006).[29]

Transcending the Abyss

So, can we coexist? Moving beyond the divide may mean that we shift our line of questioning regarding our coexistence a little. Fran Mascia-Lees (2006: 9), for example, suggests a simple yet significant alteration: *should* we coexist? In other words, do we share "compelling intellectual connections" that warrant coexistence? The answer, in my view, is simple: yes. There exists a number of anthropological problems for which our understanding would be incomplete without an integrated approach: the evolution of language, human adaptability, and race, to name just a few.[30] For each of these problems, anthropologists, sociocultural and biological alike, have come to recognize the need to uncover how biological and sociocultural processes co-shape each other. To understand the origins of language, for example, we need to not only understand how it is made possible in the human species (that is, the biological prerequisites), but also the social factors underlying its development and how it is used and interpreted in context (see Gibson & Ingold 1993). Similarly, human biological adaptation does not occur devoid of context; past and present social, cultural, and political-economic forces are always at play (see Goodman & Leatherman 1998). Consider also one of our field's most insidious mistakes: the idea that biological races are real. Integrative anthropologists have been at the forefront of correcting this claim by showing how existing biological variation does not translate into discrete races (Relethford 2009), and yet, the sociocultural reality of race and racism has biological consequences for racially defined groups. In other words, race is not biology, but can *become* biology (Gravlee 2009).[31]

Accordingly, the question should not just be can or should we exist, but *how* should we co-exist? Again, some biological and sociocultural anthropologists have grappled with this question for some time. During the "science wars" era, sociocultural anthropologist Roy Rappaport (1994) argued that an adequate understanding of humanity must encompass both law and meaning. Biological anthropologist, Matt Cartmill (1994), in his luncheon address at the 1994 meeting of the American Association of Physical Anthropologists, admitted that biological anthropologists have a lot to learn from more humanistic approaches; critical theory, for example, demonstrates how knowledge acquisition is often a political process rather than one of pure discovery.[32]

Fast forward two decades and we see that anthropologists of all flavors continue to devote attention to this question. For example, in 2010, social anthropologists, Tim Ingold and Gisli Palsson organized a panel titled "Human becomings: Beyond the 'biological' and the 'social'" at the European Association of Social Anthropologists meeting to address the biological-sociocultural divide. This panel later led to the edited volume *Biosocial Becomings: Integrating Social and Biological Anthropology* (Ingold & Palsson 2013), which is aimed at "undoing the logic" that initially led to the divide between these two subfields (Ingold 2013: 12). In 2014, anthropologists Agustín Fuentes and Polly Wiessner brought together a diverse group of anthropologists for a Wenner-Gren Foundation symposium on "Integrating Anthropology: Niche Construction, Cultural Institutions, and History." This symposium culminated in a supplementary issue in *Current Anthropology* devoted to illustrating with data how biological and sociocultural phenomena and scientific and interpretive ways of knowing can come together. Fuentes and Wiessner (2016: S3) emphasize that they "made no attempt to force uniform theoretical and methodological approaches." This, I would argue, is critical to a sustainable coexistence of biological and sociocultural anthropology.

The Biological Anthropology Section (BAS) of the AAA has also played an important role in advancing an integrative anthropology. Its members "are resolute in their identities as anthropologists" (Katherine C. MacKinnon, BAS President 2016–2018, personal communication) and the section has sponsored symposia that illustrate the relevance of biological anthropology to anthropology, such as the 2013 symposium "Entangling the biological: Steps toward an integrative anthropology" organized by Katherine C. MacKinnon and Agustín Fuentes (see Mackinnon 2014). BAS also regularly co-sponsors symposia with other sections of the AAA, including Anthropology and the Environment (A&E), Society for the Anthropology of Food & Nutrition, and the Society of Psychological Anthropology, which further exemplifies the cross-subfield nature of its members' scholarship.

As part of their rejection of the sacred bundle, Segal and Yanagisako (2005: 13) contend that those who advocate for four-field holism are essentially trying to integrate "diverse anthropologies into a singular, dominant paradigm." What they are reacting to is the view held by some that the only way the four subfields can show any semblance of integrated coherence is through a shared interest in and adoption of Neo-Darwinian evolutionary theory.[33] While contemporary evolutionary theory with its attention to other forms of inheritance – epigenetic, behavioral, ecological, and cultural – provides greater space for intellectual border crossings and mixed methodologies,[34] evolutionary approaches still represent just one path toward integration, not the *only* one. Anthropology need not be nor should be guided by one single paradigm. Moreover, anthropologists from different subfields don't need to espouse the same theoretical framework or epistemological approach to have a dialogue, or for that matter, collaborate and do integrative anthropology. So, to counter Segal and Yanagisako (2015), I argue that upholding the four-field approach means embracing the theoretical and methodological pluralism that has always characterized our field. As Silverman (2005: 332) notes, we may not be able to predict when cross-field collaboration will be fruitful (Silverman 2005: 332). We therefore need to be open to such possibilities and "preserve social spaces for reinventions of our distinctive hybridity" (Lederman 2005: 73).

Although this discussion of how to fruitfully coexist pertains more broadly to biological anthropology and its relations with sociocultural anthropology, similar sentiments have been voiced concerning primatology's relationship with sociocultural anthropology and its place in anthropology. In his *Annual Review of Anthropology* piece titled, "Whither Primatology? Primates in Contemporary Anthropology," Peter Rodman wrote: "Not only is there a lack of common interest, there is a wide epistemological abyss between sociocultural anthropology and primatology" (1999:331). Rodman further states: "Although dominant themes of current anthropology are not compatible with the epistemology, theory, or methodology of primate research and interpretation, primate studies fit easily within the future of anthropology as a four-field evolutionary study of the origins of humans and human nature" (p. 11). It is this view that I seek to unravel in this book. In the chapters that follow I explore how primatology came to be a part of anthropology and highlight the value of distinctive forms of hybridity in contemporary primatology. Fitting in anthropology in more than one way, today's and tomorrow's primatology is refreshingly intradisciplinary, but also excitingly transdisciplinary. Let's see how!

Notes

1 George Stocking Jr. (1988) notes that his colleagues at the University of Chicago disparagingly referred to the American four-field approach as the "sacred bundle." Bernard Cohn (1980: 202) also refers to it as such, albeit without derision.
2 See, for example, Stocking (1974, 1988) Darnell (1982, 2001) and Barth et al. (2005).
3 This rift, as it relates to anthropology, has a long history. See, for example, Alfred Kroeber's (1935) critique of Boas' work embedded within a broader discussion of how scientific and historical anthropology should be, and Boas' (1936) response.
4 Boas (1912).
5 While this may be the preference for many, both the journal, the *American Journal of Physical Anthropology*, and the scholarly society, the American Association of Physical Anthropologists (AAPA), both of which were established by physical anthropologist Aleš Hrdlička in 1918 and 1928, respectively, still retain their original names. At the 2017 AAPA meeting, in an invited symposium on diversity in our field in which a number of presenters reflected on current day implications of our field's racist past, biological anthropologist Robin Nelson rightfully asked with derision: "we have a problem changing our society's name because we want to honor Hrdlička?" We are getting there though, albeit slowly. An ad-hoc committee was formed to examine the implications/ramifications of changing the names, and at the 2018 meetings, the Executive Committee stated its intention to conduct an online non-binding survey all members of the AAPA regarding a name change. Of those who voted (975 valid votes of 2,416 eligible voters), approximately 82 percent voted for a name change. Of those who voted in favor of a name change, 59 percent voted for the American Association of Biological Anthropologists. Full details of the survey are available here: http://physanth.org/news/results-non-binding-name-change-survey/ . The next step is to hold a formal vote of the regular members of the association in 2019.
6 By at least the 1930s Boas dropped "biological or somatic anthropology" for the more common term at the time: "physical anthropology."
7 Spencer (1981) notes that in the latter half of the nineteenth century the term "ethnology" was the preferred term to encompass what "anthropology" means to us today.

That said, when the change to "anthropology" occurred at the turn of the century, it still largely meant just cultural anthropology and archaeology. Also of note here is that at the turn of the century, anthropology – in whatever flavor was available – was still largely taught in other departments in American colleges and universities, such as psychology, philosophy, zoology, and sociology (MacCurdy 1902).

8 Citing William Adams' "The Philosophical Roots of Anthropology," Borofsky (2002) reminds us, however, that *early* British anthropology also embraced a four-field approach. So, this four-field tradition only became distinctive for American anthropology in the twentieth century.

9 Rudolf Virchow (1821–1902) was a pathologist by training, but his interests in pathological forms of the human body led him to contribute to and help develop nineteenth-century physical anthropology in Germany (Boas 1902b). One of Virchow's students, Adolf Bastian, who was trained in law, natural science, and medicine, is credited with advancing the cultural wing of nineteenth-century German anthropology (i.e., "ethnology"); in 1869 he became the editor of *Zeitschift für Ethnologie (Journal of Ethnology)*, and along with Virchow, helped organize the Berlin Society for Ethnology, Anthropology, and Prehistory (Lowie 1937; Bunzl & Penny 2003).

10 See Hicks (2013) for further discussion of Brinton's contribution to the four-field approach of American anthropology. Hicks also acknowledges the vision that Augustus Pitt-Rivers, an English archaeologist and ethnologist, had for the teaching of anthropology; a vision that was premised on a four-quadrant structure to anthropology (see diagram p. 758), and notably, one that apparently preceded Brinton's 1892 teaching of anthropology classification by a decade and Boas' address to the International Congress of Arts and Sciences (1904) by a little more than two decades.

11 It is beyond the scope of this text to explore in detail why Boas outshined his colleagues, including those more senior to him, in our memory regarding the creation of American anthropology (see Baker 2000). It's worth noting though that Boas' biocultural approach certainly makes for a prouder anthropological legacy than does Brinton's support of scientific notions of racial inferiority.

12 Similarly, biological anthropologist James Calcagno (2003:8) addresses biological anthropologists directly on the issue of the divide: "If, as a biological anthropologist, you feel you are unwillingly being pushed out of anthropology, my recommendation is simple: Quit whining, don't drop out, push back, and help satisfy your biological needs by shaping the future of anthropology in directions you deem appropriate."

13 Morell (1993) quotes Clifford Barnett, a medical anthropologist at Stanford University: "There always was a certain amount of divisiveness, but now it's like Yugoslavia" (p. 1798) and Robin Fox, an evolutionary anthropologist at Rutgers University, "I'm afraid anthropology has become like a lunatic asylum where the patients have taken over" (p.1801). Matt Cartmill, (1994: 1), a biological anthropologist, sees anthropology as "an institutionalized train wreck" caught between science and humanities. In his contribution to the December 1995 *Anthropology News* issue on science and anthropology, Eugene Anderson, an ecological anthropologist, writes "It seems to me that science and humanism have been the two wings of the anthropological bird. Without both of them, working together, the bird doesn't fly" (p.2).

14 Segal and Yanagisako (2005: 4–5) refer to the four-field unity as "strategic holism" – an effective way of representing anthropology in the pursuit of resources and public authority. See also Stocking Jr.'s 1988 paper, *Guardians of the Sacred Bundle*.

15 See anthropologist Dell Hymes's (1972) edited volume, *Reinventing Anthropology*, which, motivated by the Vietnam war, civil rights issues and anti-colonial sentiment, called for a more reflexive and advocacy-laden approach to (sociocultural) anthropology. See also Leland Donald's (1974) review of the volume in the *American Anthropologist.*

16 The Human Evolutionary Biology website states (HEB): "As a separate department, HEB has been able to expand its scope, the size of the faculty and the graduate program, but retains ties to the Department of Anthropology and is still housed in the Peabody Museum" https://heb.fas.harvard.edu/what-is-heb.

17 A quick perusal of the department's website (https://anthropology.stanford.edu/) indicates that the four field model only really applies to the undergraduate curriculum. The graduate program requires that students select one of two tracks: culture and society, and archaeology.

18 Stocking (1960) also provides an interesting account of the apparent discord among the founding members, Franz Boas and W.J. McGee, regarding the inclusivity of the organization. Boas preferred to limit its membership to academic anthropologists, of which there were notably few at that time, while McGee's vision was for a more open organization that included practicing anthropologists as well.

19 See Eric Wolf's (1980) commentary in the *New York Times*, titled "They divide and subdivide, and call it anthropology," in which he discusses this trend toward increasing specialization.

20 The term "science" was reinstated as part of the AAA's "What is Anthropology" statement. For in-depth coverage of the removal of the term and responses to it, both inside and outside of anthropology, see Daniel Lende's 12/1/10 and 12/10/10 blog posts for "Neuroanthropology: Understanding the Encultured Brain and Body."

21 The American Association of Physical Anthropology (AAPA) was one of the "quadrant societies" that opted to not merge with the AAA during the reorganization. And, although the AAA consequently created the Biological Anthropology Section to maintain its four-field commitment, once the AAPA became a truly independent organization, few biological anthropologists attended the AAA meetings (Linda Wolfe, personal communication). See also Wolfe's (2001) discussion of the negative impact of this "desertion" on the field of biological anthropology. Interestingly, Little (2010) notes that Boas was not initially in favor of the formation of the AAPA because he was concerned that it would mean a smaller biological anthropology membership for the AAA; "history has [indeed] shown this to be true" (p.72; [my insertion]). See membership numbers in the AAA's Biological Anthropology Section below.

22 Interestingly, though, almost paradoxically, subfield-based Associate Editors were eliminated in 1990 under Keller's leadership; a structure that was not formally resurrected until Tom Boellstorff took over in 2007 (see Boellstorff 2008).

23 During this period, there were also 72 papers for which it was impossible to specify the primary subfield, so it is possible that biological anthropology was partially represented among these.

24 In 2002, the Biological Anthropology Section (BAS) membership totaled 416 (Mascia-Lees 2006); it dropped to 344 in 2015 (4.9%), rose slightly in 2016 in numbers but not in percentage of total membership (n= 355, 3.6%, and as of June 2018 is 266 (3%) (Katherine C. MacKinnnon, personal communication).

25 As of the writing of this book, the *American Anthropologist* is not indexed in Medline.

26 The survey involved telephone interviews with a random sample of 401 members (Moses 1997). The percentage breakdown of members surveyed by subfield roughly

matched the percentages represented by the respective subfields of the AAA's total membership at the time: 70% sociocultural anthropologists, 14% archaeologists, 4% linguistic anthropologists, and 7% biological anthropologists. Since 7% surpasses the number reported by Masica-Lees (2006), it is possible that the percentage includes biological anthropologists who are members of other sections (e.g., Evolutionary Anthropology) but not of BAS.

27 Here I am referring to some of the language used by Virgina Morell (1993) in her highly visible (because it's in *Science*!) piece titled "Anthropology: Nature-Culture Battleground." Although the essay conveys a positive vibe regarding anthropology's ability to transcend the divide, or the "battleground" as she calls it, I am nonetheless struck by the implicit ways blame is laid via language that disparages cultural anthropologists as, for example, being "caught up in the wave of deconstructionist thinking that has been sweeping the humanities" while victimizing biological anthropologists as having to escape into biology departments and applauding them for incorporating "the latest tools of molecular biology…" (p. 1798).

28 Lederman (2005) notes that there are also fissures internal to sociocultural anthropology; ones that follow a similar fault line as the broader field (i.e., the divide between positivist and interpretive approaches).

29 Following Goodman and Leatherman (1998), the biocultural synthesis seeks to understand how social, cultural, and political-economic contexts shape human biologies.

30 See Silverman (2005) for a review of Wenner-Gren sponsored symposia during which anthropologists have identified spaces for fruitful exchange and integration across the subfields.

31 For example, psychosocial stress experienced as a result of institutionalized racism can lead to health consequences (e.g., high blood pressure). Hence, "systemic racism becomes embodied in the biology of racialized groups and individuals, and embodied inequalities reinforce a racialized understanding of human biology" (Gravlee 2009: 54).

32 A good example is anthropologist Celia Lowe's account (2004) of how the transformation of the Togean macaque from "new form" to "endemic species" was a strategic means to legitimize a Togean conservation project that would attract significant, international funding.

33 This viewpoint has a deep history. In his anthropology textbook, British anthropologist Robert R. Marret (1912: 8) wrote: "Anthropology is the child of Darwin. Darwinism makes it possible. Reject the Darwinian point of view, and you must reject anthropology also." For more contemporary examples, see Rodman (1999), Borofsky (2002), and Blanchard (2006).

34 See Pigliucci (2007), Laland et al. (2015), and papers within the "Reintegrating Anthropology: From Inside out" issue of *Current Anthropology* (2016, v.57, S13). I explore an extended evolutionary approach for primatology in Chapter 4.

References

Anderson EN. 1995. Keeping the anthropological bird aloft. *Anthropology News* 36: 2.

Baker LD. 2000. Daniel Brinton's success on the road to obscurity, 1890–99. *Cultural Anthropology* 15: 394–423.

Balée W. 2016. The four-field model of Anthropology in the United States. *Amazônica-Revista de Antropologia* 1.

Barth F, Gingrich A, Parkin R, Silverman S. 2010. *One Discipline, Four Ways: British, German, French, and American Anthropology*: University of Chicago Press.

Blanchard S. 2006. Obscurantist holism versus clear-cut analysis: Will anthropology obviate the biology-culture divide? *Dialectical Anthropology* 30: 1–25.

Blount B. 2006. From the editor. *American Anthropologist* 108: 463.

Boas F. 1894. The correlation of anatomical or physiological measurements. *American Anthropologist* 7: 313–24.

Boas F. 1899. Some recent criticisms of physical anthropology. *American Anthropologist* 1: 98–106.

Boas F. 1902a. Some problems in North American archaeology. *American Journal of Archaeology* 6: 1–6.

Boas F. 1902b. RudolfVirchow's anthropological work. *Science* 16: 441–45.

Boas F. 1904. The history of anthropology. *Science* 20: 513–24.

Boas F. 1911a. *Handbook of American Indian Languages*. Cambridge: Cambridge University Press.

Boas F. 1911b. *Changes in the Bodily Form of Descendants of Immigrants*. New York: Colombia University Press.

Boas F. 1911c. *The Mind of the Primitive*. New York: Macmillan.

Boas F. 1912. Changes in the bodily form of descendants of immigrants. *American Anthropologist* 14: 530–62.

Boas F. 1932. Studies in growth. *Human Biology* 4: 307–50.

Boas F. 1935. The tempo of growth of fraternities. *Proceedings of the National Academy of Sciences* 21: 413–18.

Boas F. 1936. History and science in anthropology: A reply. *American Anthropologist* 38: 137–41.

Boas F. 1940. *Race, Language and Culture*. New York: The Free Press.

Boas F. 1974 [1898]. Fieldwork for the British Association, 1888–1897. In *The Shaping of American Anthropology 1883–1911, A Franz Boas Reader*, ed. GW Stocking Jr. New York: Basic Books.

Boellstorff T. 2008. From the Editor: Anthropological innovations. *American Anthropologist* 110: 1–3.

Borofsky R. 2002. The four subfields: Anthropologists as mythmakers. *American Anthropologist* 104: 463–80.

Bourguignon E. 1996. American anthropology: a personal view. *General Anthropology* 3: 1 7.

Brinton DG. 1892. The nomenclature and teaching of anthropology. *American Anthropologist* 5: 263–72.

Bruner EM. 2010. Revisiting the four fields, again. *Anthropology News* 51: 4.

Bunzl M, Penny HG. 2006. Introduction: Rethinking German anthropology, colonialism, and race. *Worldly Provincialism: German Anthropology in the Age of Empire*: 1–30.

Calcagno JM. 2003. Keeping biological anthropology in anthropology, and anthropology in biology. *American Anthropologist* 105: 6–15.

Cartmill M. 1994. Reinventing anthropology: American Association of Physical Anthropologists Annual Luncheon Address, April 1, 1994. *Yearbook of Physical Anthropology* 37: 1–9.

Chibnik M. 2013a. Continuity and change at *American Anthropologist*. *American Anthropologist* 115: 1–3.

Chibnik M. 2013b. Biological anthropology in *American Anthropologist*. *American Anthropologist* 115: 357–58.

Clifford J. 1986. Introduction: Partial truths. In *Writing Culture: The Poetics and Politics of Ethnography*, ed. J Clifford, GE Marcus, pp. 1–26. Berkeley: University of California Press.
Cohn BS. 1980. History and anthropology: The state of play. *Comparative Studies in Society and History* 22: 198–221.
D'Andrade R. 2000. The sad story of anthropology 1950–1999. *Cross-cultural Research* 34: 219–32.
Darnell R. 1982. Franz Boas and the development of physical anthropology in North America. *Canadian Journal of Anthropology* 3: 101–12.
Darnell R. 2001. *Invisible Genealogies: A History of Americanist Anthropology*. University of Nebraska Press.
Darnell R. 2002. Editor's introduction. In *American Anthropology, 1971–1995: Papers from the "American Anthropologist"*, ed. R Darnell, pp. 1–36. Lincoln, US: University of Nebraska Press.
Darnell R. 2009. Anthropological approaches to human nature, cultural relativism and ethnocentrism. *Anthropologica*: 187–94.
Darnell R, Gleach FW. 2002. *Celebrating a Century of the American Anthropological Association: Presidential Portraits*. University of Nebraska Press.
Donald L. 1974. General, applied and theoretical: Reinventing anthropology. Dell Hymes. *American Anthropologist* 76: 857–61.
Fuentes A, Wiessner P. 2016. Reintegrating anthropology. From inside out: An introduction to supplement 13. *Current Anthropology* 57: S3–S12.
Gibson K, Ingold T, eds. 1993. *Tools, Language, and Cognition in Human Evolution*. Cambridge: Cambridge University Press.
Goodman AH, Leatherman TL, eds. 1998. *Building a New Biocultural Synthesis: Political-Economic Perspectives on Human Biology*. Ann Arbor: University of Michigan Press.
Gravlee CG. 2009. How race becomes biology: Embodiment of social inequality. *American Journal of Physical Anthropology* 139: 47–57.
Hicks D. 2013. Four-field anthropology: Charter myths and time warps from St. Louis to Oxford. *Current Anthropology* 54: 753–63.
Hill JD. 1992. Contested pasts and the practice of anthropology. *American Anthropologist* 94: 809–15.
Hill JD. 1992. Overview. *American Anthropologist* 94: 809–15.
Hymes DH. 1974. *Reinventing Anthropology*. Vintage books New York.
Ingold T. 2001. From complementarity to obviation: On dissolving the boundaries between social and biological anthropology, archaeology, and psychology. In *Cycles of Contingency: Developmental Systems and Evolution*, ed. S Oyama, PE Griffiths, RD Gray, pp. 255–79. Cambridge, MA: The MIT Press.
Ingold T. 2013. Prospect. In *Biosocial Becomings: Integrating Social and Biological Anthropology*, ed. T Ingold, G Palsson, pp. 1–21. Cambridge: Cambridge University Press.
Ingold T, Palsson G, eds. 2013. *Biosocial Becomings: Integrating Social and Biological Anthropology*. Cambridge: Cambridge University Press.
Jaschik S. 2007. A forced anthropology merger. Inside Higher Education.
Keller JD. 1990. Editorial. *American Anthropologist* 92: 585–85.
Koentjaraningrat. 1987. Anthropology in Indonesia. *Journal of Southeast Asian Studies* 18: 217–34.
Kroeber AL. 1935. History and science in anthropology. *American Anthropologist* 37: 539–69.
Laland KN, Uller T, Feldman MW, Sterelny K, Müller GB, et al. 2015. The extended evolutionary synthesis: Its structure, assumptions and predictions. *Proc. R. Soc. B* 282: 20151019.

Lederman R. 2005. Unchosen grounds: Cultivating cross-field accents for a public voice. In *Unwrapping the Sacred Bundle: Reflections on the Disciplining of Anthropology*, ed. DA Segal, SJ Yanagisako, pp. 49–77. Durham: Duke University Press.

Little MA. 2010. Franz Boas's place in American physical anthropology and its institutions. In *Histories of American Physical Anthropology in the 20th Century*, ed. MA Little, KAR Kennedy. Plymouth, UK: Lexington Books.

Lowe C. 2004. Making the monkey: How the Togean macaque went from "new form" to "endemic species" in Indonesians' conservation biology. *Cultural Anthropology* 19: 491–516.

Lowie RH. 1937. *History of Ethnological Theory*. New York: Holt, Rinehart and Winston.

MacCurdy GG. 1902. The teaching of anthropology in the United States. *Science*: 211–16.

MacKinnon KC. 2014. Contemporary biological anthropology in 2013: Integrative, connected, and relevant. *American Anthropologist* 116: 352–65.

Marett RR. 1912. *Anthropology*. New York: Henry Holt & Company.

Mascia-Lees F, Lees SH. 2002. From the Editors. *American Anthropologist* 104: 713–14.

Mascia-Lees F. 2006. Can biological and cultural anthropology coexist? *Anthropology News* 47: 9–13.

Morell V. 1993. Anthropology: Nature-culture battleground. *Science* 261: 1798–801.

Moses YT. 1997. Are four fields in our future? *Anthropology News* 38: 8–11.

Nader L. 2001. Anthropology! Distinguished Lecture--2000. *American Anthropologist* 103: 609–20.

Pigliucci M. 2007. Do we need an extended evolutionary synthesis? *Evolution* 61: 2743–49.

Rappaport RA. 1994. Humanity's evolution and anthropology's future. In *Assessing Cultural Anthropology*, ed. R Borofsky, pp. 153–67. New York: McGraw Hill, Inc.

Relethford JH. 2009. Race and global patterns of phenotypic variation. *American Journal of Physical Anthropology* 139: 16–22.

Rodman PS. 1999. Whither primatology? The place of primates in contemporary anthropology. *Annual Review of Anthropology* 28: 311–39.

Rogers SC. 2001. Anthropology in France. *Annual Review of Anthropology* 30: 481–504.

Ruff CB. 2000. A view on the science: Physical anthropology at the millenium. *American Journal of Physical Anthropology* 112: 1–3.

Rutherford J. 2013. American anthropologist's "particular problem" with biological anthropology. *Anthropology News* 54: online.

Segal DA, Yanagisako SJ. 2005. Unwrapping the sacred bundle. *Reflections on the Disciplining of Anthropology.* Durham and London: Duke University Press.

Shenk MK. 2006. Models for the future of anthropology. *Anthropology News* 47: 6–7.

Silverman S. 2005. *The Boasians and the Invention of Cultural Anthropology*. pp. 257–74: Chicago: University of Chicago Press.

Spencer F. 1981. The rise of academic physical anthropology in the United States (1880–1980): A historical overview. *American Journal of Physical Anthropology* 56: 353–64.

Stocking Jr. GW. 1960. Franz Boas and the founding of the American Anthropological Association. *American Anthropologist* 62: 1–17.

Stocking Jr. GW. 1968. *Race, Culture, and Evolution: Essays in the History of Anthropology.* Chicago: University of Chicago Press.

Stocking Jr. GW. 1974. *The Shaping of American Anthropology 1883–1911: A Franz Boas Reader*. Chicago: University of Chicago Press.

Stocking Jr. GW. 1988. Guardians of the sacred bundle: the American Anthropological Association and the representation of holistic anthropology. *SB Cohen, D. Bromwich, and GW Stocking, Jr., Learned Societies and the Evolution of the Disciplines. ACLS Occasional Paper*: 17–25.

Sussman RW. 1998. Editorial. *American Anthropologist* 100: 605–06.

Tedlock B, Tedlock D. 1994. From the Editors. *American Anthropologist* 96: 521–22.

Washburn S. 1951. The new physical anthropology. *Transactions of the New York Academy of Science* 13: 298–304.

Wolf E. 1964. *Anthropology: Humanistic Scholarship in America.* Englewood Cliffs, NJ: Prentice-Hall.

Wolf E. 1980. They divide and subdivide, and call it anthropology. *The New York Times*, November 30, 1980: 4.

Wolfe LD. 2001. Biological anthropology section. *Anthropology News* 42: 46–47.

Yanagisako SJ. 2005. Flexible disciplinarity: Beyond the Americanist tradition. In *Unwrapping the Sacred Bundle: Reflections on the Disciplining of Anthropology*, ed. DA Segal, SJ Yanagisako, pp. 78–98. Durham, NC: Duke University Press.

Yanagisako SJ, Segal DA. 2006. Welcoming debate: Exploring links and disconnects among the quadrants. *Anthropology News* 47: 8–11.

3

PRIMATOLOGY

Becoming Anthropology

I am a primatologist, but who am I? Or in other words, what *kind* of primatologist am I? What lies at the heart of this question is the fact that primatology has a multidisciplinary parentage such that, today, primatologists work within a number of disciplines and academic departments, including ecology, psychology, zoology, and anthropology.[1] Being less than a century old, primatology, compared to these other "parent" fields, is a baby. The history of its development is, nonetheless, rich and complex. Because others have carefully and thoroughly captured this fascinating history, here I am admittedly selective in my coverage.[2] I have chosen to focus on the people whose works demonstrate how primatology became anthropology, and *is* anthropology. And, even with this objective I am selective in emphasizing the major developments and individuals that I find to be the most interesting or the most significant for how I envision contemporary field primatology contributing to the New Biological Anthropology (Fuentes 2010).

Additionally, as I note in Chapter 1, my coverage here focuses primarily on American primatology. Primatology emerged independently in Japan approximately the same time that it did in the West – with little exchange during the formative years (Asquith 1996: 241). Early primate studies in Japan, which began in 1948, employed a methodology derived from animal ecology (Sugiyama 1965). Although multiple schools came to comprise the "Japanese tradition of primatology," it is primarily led by scholars from the Kyoto school; an approach, founded by ecologist and anthropologist Kinji Imanishi, that is characterized by "strategic anthropomorphism, a structural view of primate societies, and highly valued long-term fieldwork" (Takasaki 2000: 163). Accordingly, most primatologists in Japan study primates from an anthropological or comparative psychology perspective, and rarely from a behavioral ecology or evolutionary biology perspective (Kutsukake 2010), which has dominated the practice of primatology in the

West since the 1970s and 1980s (see chapter 4). In Vietnam, primatologists are trained in vertebrate zoology and biology departments rather than anthropology departments; the result, according to primatologist Thach Mai Hoàng (2016), is an institutional structure that produces scholars who are unprepared to deal with the current realities of doing primate conservation with people in mind.[3]

In my above listing of primatology's diverse parentage, I purposely placed anthropology last because prior to the 1940s, there was little interest in primate behavior within the discipline of anthropology. In other words, from the perspective of anthropologists at that time, primatologists were others. In his *Man's Poor Relations*, Earnest Hooton (1942), an American anthropologist who advanced the study of biological anthropology at Harvard University, aimed to compile the "splendid research work of primatologists" (p. viii). Here he was referring to the work by C.R. Carpenter, Adolph Schultz, S. Zuckerman, and Robert M. Yerkes – all eminent scholars, but not anthropologists. And indeed this was the case, as the early years of primatology were dominated primarily by psychologists, zoologists, and ethologists. So, to understand how primatology acquired a home in anthropology, we must first meet some of the key movers and shakers of primatology's early years.

Early Movers and Shakers

Robert Mearns Yerkes (1876–1956), to whom Hooton later dedicated his 1942 book with the inscription "Student of primates; teacher of primatologists," deserves the first spotlight. Yerkes (1876–1956), a psychologist at Yale University, along with his wife, Ada W. Yerkes, undertook a substantial undertaking by compiling everything known about apes from captive studies, including his own work, and accounts from encounters in the wild (Yerkes & Yerkes 1929). Although he was primarily interested in acquiring wild caught apes for use in laboratory research on primate learning and intelligence (Ribnick 1982), Yerkes also recognized that there was a dearth of information on wild ape populations. To fill this gap in knowledge, Yerkes initiated a program for the "naturalistic study" of primates (Yerkes in Bingam 1932: 1), with the first of a series of field studies focused on two of the African great apes, chimpanzees (Nissen 1931) and gorillas (Bingham 1932). Upon publication of the studies, it became clear that Yerkes was one of the earliest scholars to appreciate what the study of primates could tell us about human origins: "intimate knowledge of this remarkable creature [the gorilla] may significantly advance our acquaintance with human origins, development, and achievements." (Yerkes 1932: 2, in the foreword to Bingham's 1932 "Gorillas in a Native Habitat).[4]

Another important early pioneer of field primatology, also trained in psychology, was Clarence Ray Carpenter (1905–1975). Upon completing his Ph.D. in Psychology at Stanford University in 1931, Carpenter joined Yerkes' lab at Yale University as a research fellow. He then became the third of Yerkes's field workers to conduct a naturalistic study, this time on a Neotropical primate. Although important first steps in the naturalistic study of primates, Nissen's and Bingham's studies were actually of short duration and minimally successful in truly expanding the scientific knowledge of primate behavior at the time (Sussman 2011). In contrast, Carpenter (1934) spent multiple years in the field conducting observations on a whole suite of behavioral and ecological parameters of howler monkeys (*Alouatta palliata*) on Barro Colorado Island, Panama. Moreover, his contributions to field primatology did not cease following this initial study; rather, he embarked on additional field studies, including observations of spider monkeys (*Ateles geoffroyi*) in Panama and a field study of gibbons (*Hylobates lar*) as part of the Asiatic Primate Expedition of 1937.

Over the course of his career, Carpenter made many contributions to primatology, but there are three that I would like to highlight here. First, Carpenter demonstrated the importance of spending significant periods of time in the field conducting observations, thereby setting the mark of what could be considered "a major field study" (Hall 1968: 9).[5] An emphasis on studies of sustained duration continues to characterize field primatology today. Second, Carpenter was committed to securing a space for field primatology in the realm of valid scientific research (Montgomery 2005). He lamented that because observational studies were based primarily on description, they were inherently subjective (Carpenter 1964).[6] Carpenter was therefore instrumental in developing new methods that would enable systematic collection of behavioral data and an experimental design to research. For example, Carpenter (1934 [1964]) developed a systematic approach for censusing that involved counting individuals as the group moved in a single file form, and an approach to the study of group coordination and integration by focusing on the dyadic relationships of different age/sex classes. He also advanced the use of technology in the field by using recording and playback equipment to "more completely and objectively [study] the vocalizations of the gibbons" and to get at the function of behavior (p. 246; see also Montgomery 2005). Lastly, Carpenter advocated for the study of primates in natural settings – that is, settings undisturbed by humans and to which the species are "fittingly adapted" (Carpenter 1942: 248). Only such studies could reveal the natural behavior of primates. Accordingly, he developed a new method to do so: through "neutral conditioning of the animal to the observer" (Carpenter 1934 [1964]: 19) or what we now call "habituation." What is interesting though is how, as Montgomery (2005) illuminates, Carpenter's notion of what counted as "natural behavior" seemed to change over time, particularly as he began to use more interventionist techniques in his studies, such as removing the alpha male from a group of rhesus macaques (*Macaca mulatta*) on Cayo Santiago Island to examine the impact on the dominance hierarchy. It is paradoxical that Carpenter didn't see such interventionist practices as an issue given that he also lamented the loss of areas where primates could be studied undisturbed by humans (see Carpenter 1950: 228). Moreover, although today habituation is seen as one of the hallmarks of field primatology, a

FIGURE 3.1 Nonhuman primates (left, a habituated group of moor macaques, *Macaca maura*) and human primates (right, researchers Kristen S. Morrow and Hendra) observing each other. Bantimurung-Bulusaraung National Park, South Sulawesi, Indonesia. Photo credit: Erin P. Riley.

necessary first step for successful fieldwork, little attention has been paid to habituation as a relational and mutually modifying process that occurs between human observers and their primate study subjects (Figure 3.1). The training we (field primatologists) acquire make clear that being ignored, or perceived as neutral, by one's study subjects is the ultimate goal of habituation. But, does our acceptance as neutral therefore mean we become a natural element of that environment? And, are we ever truly neutral elements? Having a home in anthropology enables us to explore this aspect of the human-primate interface (see Chapter 6).

Anthropology Catches On: The Founding Fathers and Mothers

Because the earliest studies of primate behavior were not conducted by anthropologists, it would not be terribly unfair to claim that anthropology was slow to catch on. Alfred Kroeber (1876–1960), Boas' first student who went on to become an eminent sociocultural anthropologist (and to father, Ursula LeGuin, for our science fiction enthusiasts), was astonished by the fact that no major study already completed on the great apes had been done by an anthropologist.[7] Kroeber (1928) was particularly interested in what the study of apes could tell us about the origin of culture, and argued for the importance of continued study of them, both experimentally and via critical observation. Margaret Mead (1942), another notable sociocultural anthropologist, praised Carpenter's field studies, not only for the insights provided on human social life, but also for the methodological guidance they provided for the study of traditional peoples.[8] And yet, decades passed before primatology acquired a solid grounding in anthropology.

It seems that scholars were reluctant to include primates in their scope. For example, to Earnest Hooton (1887–1954) it seemed natural that primates could provide a window into humanity, and so, in 1942, as Yerkes did years prior, he compiled an entire volume on primate anatomy, behavior, and biology, titling it *Man's Poor Relations*. A sad title, to say the least, but nonetheless, foundational. It is interesting, however, and for that matter, telling, that despite this effort Hooton was concerned that he might appear as a traitor to anthropology.[9] In Chapter 12, titled "Sex and Society, Antics and Semantics," Hooton wrote "I may be considered an apostate from anthropology when I assert that it seems to me that the study of individual and group behavior among the infrahuman primates may offer nearly as much to the student of man as does the investigation of the social life and psychology of contemporary savages" (p. 323). There may have been good reason for his reticence since other anthropologists were less convinced about what primates could tell us about human social life. For example, sociocultural anthropologist Marshall Sahlins (1960) saw primate sociality as driven by sexual attraction and human social life as culturally determined, thereby pitting biology against culture. Sahlins was therefore unable to see or appreciate interesting examples of social learning and its importance for nonhuman primates.[10] A little more than a decade after his 1942 volume, and just prior to his death, Hooton was much more confident in his support for the contribution of primatology for anthropology. With this confidence in stride, Hooton (1954) laid out his vision for what the study of primates could tell us about being human, thereby creating a home for primatology within American anthropology.

Hooton titled his piece "The Importance of Primate Studies in Anthropology." Let's pause for a moment and think about this title. Why do academics write papers with a title that approximates something like "the importance of X for Y"? They do so when they sense that a particular topic or field is underappreciated, and needs greater attention. This is precisely what Earnest Hooton did with his 1954 paper, and, it worked. In this landmark piece, Hooton (1954) explicitly and comprehensively outlined what he envisioned as the specific contributions of primatology to the field of anthropology, including, how the comparative study of anatomy and physiology of the primates expands our understanding of biological taxonomy and human evolution; the value of experimental research on physiology and pathology for humanity; the implications of primate social life for human social origins and the origin of culture; and the value of studies on personality, temperament, and intelligence for our understanding of human mental development. Although Hooton's primary interest was in anatomy, he also fueled interest in behavior, and given that his vision was about situating primatology in anthropology, Hooton was particularly interested in what the study of primates could tell us about the beginnings of social organization, and even argued that the study of primates "living under natural conditions" could tell us more about this topic than could studies on contemporary traditional peoples (p.185). This vision led to Hooton, along with his student Sherwood ("Sherry") Washburn, stressing the need for primate *field* research, thereby establishing the tradition of field primatology in American physical (biological) anthropology.

If Hooton was the visionary, then Washburn was the prime mover (Ribnick 1982). Simply put: Washburn articulated the purpose of anthropological primatology, garnered the interest and funds to support it, and produced an impressive array of students who made it happen. He was *the* pioneer of field primatology from an anthropological perspective. When setting the scope of anthropological primatology, Hooton (1954: 186) himself lamented "the lack of sustained interest in such [primate] studies, the present reluctance of institutions to support them, and the difficulty in financing them."[11] Washburn changed that. Washburn clearly had strong networking skills through which he was able to garner considerable interest in the study of primate behavior, which ultimately led to research funding available. For example, Washburn befriended Paul Fejos, the founder and director of the Viking Fund (now the Wenner-Gren Foundation for Anthropological Research), who became a major supporter of primate field research and the dissemination of biological anthropology via a series of summer seminars (Howell 2003; Tuttle 2000). In fact, it was Fejos who was responsible for funding Washburn's first trip to Africa in the spring of 1948 to examine australopithecine fossils in South Africa and to collect and observe several cercopithecine taxa (Washburn 1983; Howell 2003).

Between the years of 1929–1960, only two anthropology dissertations were completed on primates; both focused on anatomy, and one of these was Washburn's (Gilmore 1981). His dissertation involved a metrical analysis of the skeletal proportions of adult langurs and macaques (Washburn 1942). Washburn was trained as an anatomist, so it is not surprising that he was influenced by the prevailing emphasis on morphology and taxonomy in biological anthropology (Stini 2010).[12] And, in fact, his first academic position was in anatomy at Columbia University (1939–1947). But Washburn was also paying attention to what was going on more broadly in evolutionary biology;[13] namely, the articulation of Charles Darwin's and Alfred Wallace's theory of evolution by natural selection with Mendelian genetics, or the "Modern Synthesis." Not only did the modern synthesis provide the fodder for Washburn's vision for the New Physical Anthropology, which we discussed earlier in Chapter 1, but it also indirectly opened the door for primate field studies due to the recognized importance of behavior in the evolutionary process (Ribnick 1982). In essence, Washburn recognized the importance of triangulation – that no one source of data alone could provide all the answers to human origins. As he saw it, fossil evidence provided the physical evidence to chart the course of evolution, genetics provided critical information about the process of evolution, and successful behaviors exemplified natural selection at work (Washburn 1968). Over the years, then, we see that Washburn's interest and focus shifted away from strict anatomy (and the use of a static typological approach of many of his mentors, including his own advisor, Hooton) toward what Stini (2010: 177) refers to as an "amalgamation of functional anatomy, population genetics, and behavioral biology."[14]

Interestingly, though, Washburn apparently did not initially take to the study of behavior. During his first field expedition, the 1937 Asiatic Primate Expedition in Thailand, Washburn spent much more time with Adolph Schultz, the primate morphologist on the team, assisting with the collection and dissection of primates, than he did with Carpenter, whose focus was the social behavior of gibbons (Washburn 1983). Tuttle (2000: 865) notes that retrospectively Schultz "was amused that in 1937 Sherry saw little to be gained from Carpenteresque studies of wild primates."[15] So, it wasn't really until the 1950s that Washburn was on board. Washburn's broadened perspective was undoubtedly the result, at least in part, of Hooton's support for primate field studies in anthropology. But, two additional factors were instrumental. In 1955, Washburn made his second trip to Africa, this time to Zambia and Zimbabwe. It was during this trip that he came to regard the study of behavior as rewarding, so much so that he "closed out the collecting and spent my time watching the tame baboons" (Washburn 1983: 16). Perhaps it was because of the expanded duration of his observations, or because the baboons were habituated to people, enabling him to get up close and personal and see, for example, that "almost at once the animals ceased to be just baboons; they became personalities" (Washburn 1983: 16). Regardless, his experiences during this trip led him to appreciate what the comparative study of primate social behavior could tell us about human evolution. It is also possible that his broadened perspective coincided with his taking his first anthropology position in the academy. In 1947, Washburn joined the anthropology department at the University of Chicago, where he was influenced by the strong tradition of social anthropology of the department.[16]

Indeed, Washburn was the prime mover in establishing anthropological field primatology, but as noted above, Washburn himself did very little primate field research. It was others, primarily his students, who led the charge in terms of fieldwork and advancing anthropological field primatology during the formative years. Washburn's student, Phyllis Jay (now Dolhinow), in particular, deserves recognition for emphasizing the complexity of primate behavior and for coalescing the growing scientific knowledge about primate behavior and ecology and its relevance for anthropology through her edited volumes: *Primates: Studies in Adaptation and Variability* (1968) and *Primate Patterns* (1972). Dolhinow is inarguably a founding mother of anthropological field primatology. I would be remiss if I did not also acknowledge the contributions of the other founding mothers of field primatology: Jane Goodall, Dian Fossey, and Birute Galdikas.[17] Although it was only Galdikas who had training in anthropology, all three of them produced, through their long-term fieldwork involving in-depth immersion into the lives of individual great apes, critical insights into great ape societies that rocked anthropology's world: observations of infanticide among mountain gorillas (Fossey 1984), documentation of the slow histories of orangutans (Galdikas 1981), and observations of tool use behavior that forced anthropologists to reconsider what it means to be human (Goodall 1964).

Primatology and the "New Physical Anthropology"

Prior to Hooton and Washburn, the focus of primatological research in anthropology was largely anatomical. As part of his "The New Physical Anthropology" Washburn called for the move away from the traditional method of comparing anatomical structures toward a new one focused on adaptive structures and explaining the function of structures. Washburn (1963) argued that to determine the function, one had to understand how the parts work; in other words, one had to examine the behavior of the living animal, preferably under natural conditions. Moreover, he argued that by looking at how living primates use structures today, we can then begin to piece together how our ancestors might have lived. Washburn recognized that for such gregarious animals as primates, an understanding of the social context in which primate bodies and anatomies exist was critical to understanding adaptations. This focus was nicely illustrated by Washburn in his 1967 *Huxley Memorial Lecture* during which he used the patas monkey (*Erythrocebus patas*), the cheetah of the primate order, if you will, as an example. He noted that although the fast running ability is critical to the success of the patas monkey as a ground-dwelling primate, "this locomotor adaptation is successful only because it is part of a particular social system" (Washburn 1967: 21); that is, differential behavior exhibited by males (flee) and females (freeze in place) when confronted with a predator. Washburn's interest was, therefore, not limited to observing how anatomical structures worked, but also on what the behavior, in particular social behavior, could tell us about the evolution of human behavior. And there you have it; primatology in *biological* anthropology is born. The biological anthropology that emerged following Washburn's call for a new approach was dynamic, comparative, evolutionary, and the other primates had a seat at the table; they could be studied as models in order to reconstruct human origins and human behavioral evolution.

From Referential Models to a Broad Comparative Analysis

Savannah-dwelling baboons (*Papio* spp.) were immediately viewed as a suitable referential model given their terrestrial lifestyle and the assumption at the time that the earliest hominins evolved in a similar environment. The idea was that by studying baboon behavior researchers could begin to surmise the constraints that our ancestors faced when they became upright and bipedal (Washburn & DeVore

1961). For example, DeVore and Washburn (1963), in their study of olive baboons (*Papio anubis*) in Kenya, envisioned early hominins relying on a similar diet mostly comprising vegetation, occasionally supplemented by meat from small animals, and exhibiting a similar habitat use pattern that revolved around access to water.[18] By observing how baboons handle life in the savannah, anthropologists could then develop testable hypotheses about the problems our early ancestors faced and test those hypotheses against the fossil record. Chimpanzees (*Pan troglodytes*), who also spend a good portion of the day on the ground, were also seen as a good model and eventually surpassed baboons as the best referent as accumulating morphological and genetic data indicated a closer phylogenetic tie with humans, and field studies documented their tremendous capacity for behavioral flexibility. Washburn (1967) had a strong interest in the locomotor adaptation of chimpanzees – knuckle-walking – and surmised that our ancestors likely locomoted in a similar manner prior to selection favoring the transition to bipedal locomotion.[19] Similarly, the social behavior of chimpanzees, along with the other African apes, could be studied in order to infer what the behavior of the last common ancestor might have been like.

Although this early modeling focused primarily on single species (i.e., the baboon model, the chimpanzee model), Washburn and other early primatologists (e.g., K.R.L. Hall, John E. Frisch) also recognized that the comparative analysis they advocated necessitated a broad purview, meaning that a complete understanding of behavioral flexibility and the similarities and dissimilarities between humans and primates could only come from studies *across* the primate order. Scholars came to recognize that no one single primate species or one model in particular could sufficiently reconstruct the behavior of early hominins; human uniqueness was just too great (Kinzey 1987).[20] Studying the behavior of the diverse array of living primates helps us understand what is possible, and hence what was probable for our early ancestors (Strum & Mitchell 1987). For example, anthropologist Shirley Strum's finding that baboons employ other, non-aggressive strategies (e.g., friendships) in competitive contexts suggests that such social strategies have a much deeper evolutionary history than previously believed (Figure 3.2). Strum and Mitchell (1987: 101) wrote:

> Knowing that "lowly" baboons are capable social strategists has important implications for how we characterize aggression among other primates, including humans. If aggression is just one strategy among several, and if it is often the least effective and riskiest option, then selection should have been for a flexible response preparing an individual for the possibility of aggression but not locking him into an aggressive reaction. Finding such flexibility among baboons changes the evolutionary timescale in a significant way. Baboon social strategies imply that no matter how ancient aggression might be, the social alternatives we observe today are certainly older than previous interpretations have suggested. Once created, these social alternatives must

FIGURE 3.2 Baboon friendship: females and their families bonded to the male being groomed, Uaso Ngiro Baboon Project, Laikipia, Kenya. Photo credit: Shirley C. Strum.

> have quickly displaced aggression from its pivotal place as *the* strategy to that of a tactic in a larger array of options. With more evolutionary time and great selective pressures for the development of flexibility, it becomes increasingly difficult to posit the inevitability and centrality of aggression among primates, including humans. (p.99, original emphasis)

Similarly, Potts (1987: 44) argued that "field studies of primates not only describe the static results of adaptation but also show *how* primates adapt and *how* behavioral variations arise" (original emphasis). In this view, all of the primates, not just baboons or chimpanzees, become important sources of insight. Studying, for example, how and why some primates exist in multilevel societies (e.g., snub-nosed monkeys, *Rhinopithecus* spp.) may shed light on how complexity emerged in modern human social organization (Grueter 2012). Elucidating the fitness benefits of forming long-term pair bonds in socially monogamous nonhuman primate species, such as the owl monkey (*Aotus* spp.; Fernandez-Duque & Huck 2013), can provide insight into the factors that led to pair bonding in humans.

In essence, simple analogies and direct comparisons between humans and our closest living relatives were traded in for an approach aimed at uncovering the key principles and processes that explain behavioral variation in an evolutionary context (Strier 2003a; 2011). And importantly, identifying those general principles is not possible through the study of just one species or even a small group of species (e.g., humans and the great apes) (cf. Martin 2002). Expanding the taxonomic breadth of the study subjects allowed primatologists to see variation across species, thereby disrupting our ideas about what is "typical" about primates. For example, in a classic paper titled "The Myth of the Typical Primate," Karen Strier (1994) argued that because early on in the development of the field terrestrial cercopithecines (e.g., baboons and macaques) were the primary study subjects, the cercopithecine model of male dispersal and female philopatry and the resulting social dynamics that ensue from this social organization came to represent the order as a whole.[21] By widening the scope to include, for example, the lemurs and Neotropical primates, primatologists recognized that there are other ways of "being a primate" – e.g., females can dominate over males, males can remain in the natal groups with their brothers, and aggressive competition can be sparse to non-existent. A broad comparative approach also enabled us to explain interspecific variation. Namely, comparative methods that control for phylogeny could be used to better understand the relative importance of phylogeny and ecology in explaining behavior; that is, determining which behaviors are conservative and which are more plastic and vary in accordance with local ecological and social conditions (Strier 2003b).

The influx of primatologists doing fieldwork, in general, and their expanded taxonomic purview, in particular, also meant that behavioral variation *within* species became visible. Strier (2003b) notes that this was made possible by studying multiple populations of the same species and increasing the duration of field studies. Importantly, a focus on the causes and consequences of *intraspecific* variation helped to rectify some of the distortions that resulted from within-species generalizations that derived from a limited number of studies or studies conducted in only one type of setting; for example, the homogenization of baboon behavior. In the early years of field primatology, studies of baboons were initially conducted on those living in open grasslands, and focused on the importance of aggression in such harsh environments, particularly among males (DeVore 1965, but see Strum 1982).[22] However, zoologist Thelma Rowell (1966) studied forest-dwelling olive baboons (*Papio anubis*) in Uganda and found that they experience a richer food supply and exhibit more relaxed and friendly groupings with adult males primarily engaging in mutual grooming with females and being the targets of play with juveniles.[23] Studying multiple populations of the same species with a focus on capturing intraspecific variation enables us to better understand how demography and life history processes affect behavior, and whether and how primates adjust their behavior in response to local ecological conditions, which in turn informs our understanding of the adaptive potential of primate species (Strier 2017).

By replacing the reconstruction of human behavioral evolution via referential models with the application of the general principles of evolutionary theory, primatology affirmed its linkages beyond its disciplinary home, most notably, to evolutionary biology. We will examine these linkages in more detail in the next chapter, but it is imperative that we remember that despite this shift, primatology's connection to anthropology remains: the broad comparative, cross-species (primate) perspective that characterizes the current scope of behavioral primatology in anthropology is an extension of the very essence of the discipline of anthropology, the cross-cultural perspective (Fedigan 2000; Pavelka 2002).

How to do Anthropological (Behavioral) Primatology?

By the late 1950s, primate field studies had taken off, and this expansion was largely the result of the work of Washburn's students; in all, Washburn produced 17 students with PhD's in field primatology (Kelley & Sussman 2007). Washburn produced so many students, and outstanding students, at that, who went on to secure academic jobs and produce students of their own, that many American anthropological primatologists today can trace their academic genealogy back to Washburn (and, of course, Hooton) (Figure 3.3).[24] We have to remember, however,

FIGURE 3.3 Three generations of primatologists who trace their academic genealogy back to Washburn and his PhD students. Photo taken at 2015 American Anthropological Association meeting in Denver, Colorado. Left to right: Nancy Krusko (earned PhD under Washburn's student Phyllis Dolhinow, and was undergraduate advisor of the author), the author, Erin Riley (earned PhD under Carolyn Ehardt), Carolyn Ehardt (earned PhD under Washburn's student, Claud Bramblett), KT Hanson (earned MA under Erin Riley), and Alison Zak (earned MA under Erin Riley). Photo credit: Katherine C. MacKinnon.

that most of the biological anthropologists at the time were trained in anatomy, and nineteenth-century typological anatomy at that (DeVore & Washburn 1992: 416). So how did they *do* anthropological primatology? Washburn's *The New Physical Anthropology* had already hit the scene, providing the rationale for the study of behavior, but still, the conceptual, theoretical, and methodological guidance for the early anthropological primatologists came not from biological anthropology per se but from British social anthropology; specifically, the prevailing framework in social anthropology in the 1950s and 1960s: functionalism. In retrospect, this influence is not surprising for at least two reasons: biological anthropologists had no training at the time in the study of behavior, human or otherwise, and the Department of Anthropology at the University of Chicago, where Washburn was faculty (1947–1958), was known for its emphasis in structural-functionalism well after Radcliffe-Brown's departure in 1937 (Gilmore 1981).

Functionalism emphasized the significance or function of phenomena (e.g., kinship, descent, individual social relationships) in maintaining society. Given his own interest in adaptive (biological) complexes, Washburn was very much impressed with Malinowski's goal of looking for the functional patterns of social phenomena (DeVore & Washburn 1992: 518). Gilmore (1981) also notes the influence of A.R. Radcliffe-Brown's flavor of functionalism, structural-functionalism, on the work of Washburn and his early students. For example, Irven DeVore (1962), Washburn's first PhD student and the first anthropologist to produce a dissertation on the behavior of free-ranging nonhuman primates, highlighted the male dominance hierarchy as the core structure of baboon society, thereby mirroring Radcliffe-Brown's emphasis on how social phenomena (e.g., customs, rituals) contribute to the reproduction of (human) society as a whole. Likewise, Phyllis Jay's (now Dolhinow) 1963 dissertation on the common Indian langur, "the most ground living of the langurs," is structured around detailed examination of each age/sex class group and their social relationships with one another, culminating in a discussion chapter that emphasizes the importance of these relationships for maintaining group cohesiveness; other aspects of behavior, such as diet and communication, are relegated to appendices.[25]

Early anthropological primatologists also turned to social anthropology for methodological guidance, despite the fact that, as we learned above, Carpenter had already pioneered a number of techniques in behavioral observation. Take Hooton's 1942 treatise on the state of knowledge of primates at the time in which he states his aim to study them as he would study man, as "primate ethnography." The pioneering studies in anthropological primatology therefore emphasized fieldwork, again reflecting Malinowski's influence, and relied upon thick, rich description of group structure, social roles, and relationships.[26]

Much has changed, in terms of theoretical origin and methodological approach, since the early years of anthropological primatology.[27] We will explore some of these changes in the next chapter, but to close this chapter I'd like to return to the idea of primatology's place in anthropology. Today primatology remains situated as one of the subfields of biological anthropology,[28] but I think it is important to emphasize that Hooton's vision was about the importance of primatology to anthropology *as a whole*, not just biological anthropology, or for that matter, sociocultural anthropology. The fact that primatology ended up being relegated to biological anthropology may have more to do with a) the fact that physical anthropologists were already immersed in comparative study of primate anatomy, b) that the study of human evolution was predominantly about humans as biological beings, and hence, the parallel to the other primates was their biological closeness, and c) that Hooton himself was a physical anthropologist. I make this point because as I argue herein, the promise of contemporary primatology is in its articulations across anthropology's subfields – as *anthropological* practice – but also beyond the disciplinary borders of anthropology itself. The increasingly predominant context in which today's primatology takes place is anthropogenic – human presence and activity play a role, either directly or indirectly, in shaping the environments in which primates live. Accordingly, contemporary primatology is about looking beyond the borders of primatology's original construction. In the chapters that follow I explore the anthropogenic context and primatology's fruitful articulations with both familiar and less-charted territories.

Notes

1 Primatologist Alison Richard (1981) surmises that because primates are "inaccessible, long-lived, behaviorally complex, and difficult to manipulate experimentally, … [they were] left to the attention of students of human nature: anthropologists and psychologists" (p. 518).
2 See Ribnick (1982); Strum & Fedigan (1999); Strier (2003a and b); Sussman (2011).
3 Chapter 7 addresses what primatologists can learn from the anthropology of conservation.
4 Or at least one of the earliest scholars to publish the idea.
5 In his survey of field studies, Hall (1968) defined "major studies" as those in which at least three months was spent working with the same species. Carpenter's (1940[1964]) three-month study of *Hylobates lar* in Thailand was the first of these. Later "long-term" studies came to mean those that spanned at least a year, i.e., to capture seasonal variation. Strier (2003b) notes that today "long-term" denotes decades of research.
6 Carpenter's view departs from those of other notable ethologists, such Niko Tinbergen (1963) and Jeanne Altmann (Altmann & Altmann 2003), who recognized the value of description and natural history for the science of ethology. In subsequent chapters I further explore the importance of description (Chapter 5) and subjectivity (Chapter 6) for field primatology.
7 Kroeber (1938:327) wrote: "Strangely enough, not one of the studies of the great apes has been made by an anthropologist."
8 Mead was impressed with Carpenter's ability to remove human projection from his data collection and analyses. She envisioned that observers of "primitive peoples" could learn from this approach by being careful about projecting their own culture onto their study subjects.

9 In line with the theme of this book, it is worth pointing out that Hooton too was an early supporter of an integrated anthropology: "if anthropology is going to amount to anything as a complete science of man, it will have to avoid splitting itself up into specialties that tend to secede and join up with other disciplines" (1954, p. 187).

10 A year later, Washburn and DeVore (1961: 62) challenged this contention (i.e., that sex is what unites a social group) by arguing, based on their field research on olive baboons, that "it is the intensely social nature of the baboons…that keeps the troop together."

11 As we shall see in Chapter 5, ethnoprimatology faced similar challenges as it emerged in the early 2000s. While more and more ethnoprimatological projects are being funded and published, the field still faces some skepticism regarding its contributions to science. It is my hope that this text will help stimulate further interest and dissolve the remaining uncertainty of its rigor and scientific impact.

12 Although Washburn majored in anthropology at the undergraduate level, he never took coursework in anthropology in graduate school (DeVore & Washburn 1992).

13 Again, likely due to his networking skills. In his "Evolution of a teacher" paper, Washburn (1983) notes that he and Theodosius Dobzhansky, a renowned evolutionary biologist, became friends immediately upon Dobshansky learning that Washburn, too, favored populations over types.

14 While he respected his former PhD advisor, Washburn (1982:2) was critical of Hooton's views on race and his research methods: "All of Hooton's major studies depended on the statistical validation of typologies and his advanced teaching was simply incorrect."

15 Washburn's recounting of it was a bit different though: "I spent three or four weeks there helping him [Carpenter], but it was just the wrong time to expect me to shift gears from anatomy to behavior…" (DeVore & Washburn 1992: 415).

16 Anthropologist Jonathan Marks (2000) notes that Washburn was supposed to have received a joint appointment in anatomy but when that fell through, anthropology gave him a full position. Without the link with the anatomy department, Washburn didn't have access to resources to conduct experimental research, and hence, shifted his focus to the study of primate behavior. How fortuitous for us anthropological primatologists!

17 And, of course, Louis Leakey, the British/Kenyan paleoanthropologist who garnered initial funding for their fieldwork (see Morell 1993).

18 It is important to point out here that Washburn and DeVore (1961: 71) also recognized the potential for significant differences between baboons and early hominins. They make this clear by charting "the salient features of baboon social life in a native habitat alongside of those of human life in pre-agricultural society" (e.g., social control, social organization, sexual behavior, etc.).

19 This hypothesis was eventually abandoned as the expanding hominin fossil record provided skeletal evidence (e.g., *Ardipthecus*) that the last common ancestor of humans and the African apes was decidedly not a knuckle-walker (see Sayers et al. 2012 for a review).

20 Twenty-four years after Washburn's Wenner-Gren conference "Social Life of Early Man" in 1959 in which he argued for the role primate field studies could play in reconstructing early hominin behavior, a group of scholars convened at the American Anthropological Association meetings to, yet again, consider "why should anthropologists be concerned about nonhuman primates?" (Kinzey 1987: vii).

21 See also Strier (2001).

22 I explore this further in Chapter 6.

23 See also Rowell (2000) for a critique on such generalizations.

24 In their 2007 review, Kelley and Sussman found that approximately 61 percent of recorded field primatologists are linked to the Hooton/Washburn lineage. See also the online resource for the academic phylogeny of biological anthropology at www.physanthphylogeny.org/FAQ.

25 I use DeVore and Jay's work as examples, but as noted by Gilmore (1981), the structural-functionalist approach pervades many of the early primate field studies.

26 For a taste of this rich, descriptive approach: "Although the young infant-2 is away from the mother for increasingly longer periods, she is alert to its whereabouts. If the troop is alerted, disrupted, or moved suddenly, the young infant either runs to her or she runs to it. The infant sits by her side and grasps her fur until the danger or tension ceases and it can resume its relaxed play. At times the infant may wander as far as one hundred feet from its mother, especially when it is following play groups of other small infants. Usually, however, these plays groups and clusters of small infants playing separately but near each other are close groups of adult females. The young infant-2 does not travel with a group of infants when the troop moves; instead it always returns to its mother and it is carried. In general it spends no more than about 45 minutes away from the mother at one time and is usually within sight of her" (Jay 1963: 43).

27 For a detailed history of the stages in development of the theory and method of primatology see Strum and Fedigan (1999).

28 And, no less important is the fact that in the U.S. most primatologists find academic jobs in anthropology departments.

References

Altmann SA, Altmann J. 2003. The transformation of behaviour field studies. *Animal Behaviour* 65: 413–23.

Asquith PJ. 1996. Japanese science and western hegemonies: primatology and the limits set to questions. In *Naked Science: Anthropological Inquiry Into Boundaries, Power, and Knowledge*, ed. L Nader, pp. 239–58: Routledge.

Bingham HC. 1932. *Gorillas in a Native Habitat*. Oxford: Carnegie Institution.

Carpenter CR. 1934. A field study of the behavior and social relations of howling monkeys (*Alouatta palliata*). *Comparative Psychology Monographs* 10: 1–168.

Carpenter CR. 1942. Section of anthropology: characteristics of social behavior in non-human primates. *Transactions of the New York Academy of Sciences* 4: 248–58.

Carpenter CR. 1950. Social behavior of non-human primates. *Colloques Internationaux* 34: 227–46.

Carpenter CR. 1964. *Naturalistic Behavior of Nonhuman Primates*. University Park, PA: Pennsylvania State University Press.

DeVore I. 1962. *The Social Behavior and Organization of Baboon Troops*. Chicago, IL: University of Chicago.

DeVore, I. 1965 Male dominance and mating behavior in baboons. In *Sex and Behavior*, ed. FA Beach, pp. 266–289. New York: John Wiley & Sons.

DeVore I, Washburn SL. 1963. Baboon ecology and human evolution. In *African Ecology and Human Evolution*, ed. FC Howell, F Bourliere, pp. 335–67. Chicago: Aldine.

DeVore I, Washburn SL. 1992. An interview with Sherwood Washburn. *Current Anthropology* 33: 411–23.

Dolhinow P. 1972. *Primate Patterns*: Holt, Rinehart & Winston.

Fedigan LM. 2000. News and views. *American Journal of Physical Anthropology* 113: 451–54.

Fernandez-Duque E, Huck M. 2013. Till death (or an intruder) do us part: intrasexual-competition in a monogamous primate. *PloS one* 8: e53724.

Fossey D. 1984. Infanticide in mountain gorillas (*Gorilla gorilla beringei*) with comparative notes on chimpanzees. In *Infanticide: Comparative and Evolutionary Perspectives*, ed. G Hausfater, SB Hrdy, pp. 217–36. New York: Aldine.

Fuentes A. 2010. The new biological anthropology: Bringing Washburn's New Physical Anthropology into 2010 and beyond--The 2008 AAPA luncheon lecture. *Yearbook of Physical Anthropology* 53: 2–12.

Galdikas BMF. 1981. Orangutan reproduction in the wild. In *Reproductive Biology of the Great Apes: Comparative and Biomedical Perspectives*, ed. C Graham, pp. 281–99. New York: Academic Press.

Gilmore H. 1981. From Radcliffe-Brown to sociobiology: Some aspects of the rise of primatology within physical anthropology. *American Journal of Physical Anthropology* 56: 387–92.

Goodall J. 1964. Tool-using and aimed throwing in a community of free-living chimpanzees. *Nature* 201: 1264–1266.

Grueter CC. 2012. Is there anything to learn from lesser known primates about human evolution? *Journal of Primatology* 1: e114.

Hall KRL. 1968. Social organization of the Old-World monkeys and apes. In *Primates: Studies in Adaptation and Variability*, ed. PC Jay, pp. 7–30. New York: Holt, Rinehart, and Winston.

Hoàng TM. 2016. Development of primatology and primate conservation in vietnam: challenges and prospects. *American Anthropologist* 118: 130–37.

Hooton EA. 1942. *Man's Poor Relations*. Doubleday, Doran.

Hooton EA. 1954. The importance of primate studies in anthropology. *Human Biology* 26: 179–88.

Howell FC. 2003. Sherwood Larned Washburn (1911–2000). *Biographical Memoirs* 84: 1–24.

Jay PC, ed. 1968. *Primates: Studies in Adaptation and Variability*.New York: Holt, Rinehart and Winston.

Jay PC. 1963. *The Social Behavior of the Langur Monkey*. Chicago: University of Chicago.

Kelley EA, Sussman RW. 2007. An academic genealogy on the history of American field primatologists. *American Journal of Physical Anthropology* 132: 406–25.

Kinzey WG. 1987. Introduction. In *The Evolution of Human Behavior: Primate Models*, ed. WG Kinzey, pp. vii–xvi. Albany, NY: State University of New York Press.

Kroeber AL. 1928. Sub-human culture beginnings. *The Quarterly Review of Biology* 3: 3225–342.

Kutsukake N. 2010. Lost in translation: Field primatology, culture, and interdisciplinary approaches In *Centralizing Fieldwork: Critical Perspectives From Primatology, Biological Anthropology and Social Anthropology*, ed. J MacClancy, A Fuentes, pp. 104–20. New York: Berghahn Books.

Marks J. 2000. Sherwood Washburn, 1911–2000. *Evolutionary Anthropology: Issues, News, and Reviews* 9: 225–26.

Martin RD. 2002. Primatology as an essential basis for biological anthropology. *Evolutionary Anthropology: Issues, News, and Reviews* 11: 3–6.

Mead M. 1942. Review of "A Field Study in Siam of the Behavior and Social Relations of the Gibbon" by R.C. Carpenter. *American Anthropologist* 44: 511–12.

Montgomery GM. 2005. Place, practice and primatology: Clarence Ray Carpenter, primate communication and the development of field methodology. *Journal of the History of Biology* 38: 495–533.

Morell V. 1993. Called "Trimates," three bold women shaped their field. *Nature* 260: 420–25.

Nissen HW. 1931. *A Field Study of the Chimpanzee: Observations of Chimpanzee Behavior and Environment in Western French Guinea.* Baltimore: Johns Hopkins Press.

Pavelka MM. 2002. Resistance to the cross-species perspective in anthropology. In *Primates Face to Face: The Conservation Implications of Human-Nonhuman Primate Interconnections*, ed. A Fuentes, L Wolfe, pp. 25–44.

Potts R. 1987. Transportation of resources: Reconstructions of early hominid socioecology: a critique of primate models. In *The Evolution of Human Behavior: Primate Models*, ed. WG Kinzey, pp. 28–46. Albany, NY: State University of New York Press.

Ribnick R. 1982. A short history of primate field studies: Old World monkeys and apes. In *A History of American Physical Anthropology 1930–1980*, ed. F Spencer, pp. 49–73. New York: Academic Press.

Richard AF. 1981. Changing assumptions in primate ecology. *American Anthropologist* 83: 517–33.

Rowell T. 2000. A few peculiar primates. In *Primate Encounters: Models of Science, Gender, and Society*, ed. SC Strum, L Fedigan, pp. 57–70. Chicago: University of Chicago Press.

Rowell TE. 1966. Forest living baboons in Uganda. *Journal of Zoology* 149: 344–64.

Sahlins MD. 1960. The origin of society. *Scientific American* 203: 76–86.

Sayers K, Raghanti MA, Lovejoy CO. 2012. Human evolution and the chimpanzee referential doctrine. *Annual Review of Anthropology* 41: 119–38.

Stini WA. 2010. Sherwood L. Washburn and "The New Physical Anthropology". In *Histories of American Physical Anthropology in the 21st Century*, ed. MA Little, KAR Kennedy, pp. 173–85. Lanham: Rowman & Littlefield Publishers, Inc.

Strier KB. 1994. Myth of the typical primate. *Yearbook of Physical Anthropology* 37: 233–71.

Strier KB. 2001. Beyond the apes: Reasons to consider the entire primate order. In *Tree of Origin: What Primate Behavior Can Tell Us about Human Evolution*, ed. FBM de Waal, pp. 69–93. Cambridge: Harvard University Press.

Strier KB. 2003a. Primate behavioral ecology: From ethnography to ethology and back. *American Anthropologist* 105: 16–27.

Strier KB. 2003b. Primatology comes of age: 2002 AAPA Luncheon address. *American Journal of Physical Anthropology* 122: 2–13.

Strier KB. 2011. Why anthropology needs primatology. *General Anthropology* 18: 6–8.

Strier KB. 2017. What does variation in primate behavior mean? *American Journal of Physical Anthropology* 162: 4–14.

Strum SC. 1982. Agonistic dominance in male baboons: An alternative view. *International Journal of Primatology* 3: 175–202.

Strum SC, Fedigan L. 1999. Theory, method, gender, and culture: What changed our views of primate society? In *The New Physical Anthropology: Science, Humanism, and Critical Reflection*, ed. SC Strum, DG Lindburg, D Hamburg, pp. 67–105. Upper Saddle River, NJ: Prentice Hall.

Strum SC, Mitchell W. 1987. Baboon models and muddles. In *The Evolution of Human Behavior: Primate Models*, ed. WG Kinzey. Albany: State University of New York Press.

Sugiyama Y. 1965. Short history of the ecological and sociological studies of non-human primates in Japan. *Primates* 6: 457–60.

Sussman RW. 2011. A brief history of primate field studies. In *Primates in Perspective*, ed. CJ Campbell, A Fuentes, KC MacKinnon, SK Bearder, RM Stumpf. Oxford: Oxford University Press.

Takasaki H. 2000. Traditions of the Kyoto School of field primatology in Japan. In *Primate Encounters: Models of Science, Gender, and Society*, ed. SC Strum, L Fedigan, pp. 151–64.

Tinbergen N. 1963. On aims and methods of ethology. *Zeitschrift für Tierpsychologie* 20: 410–33.

Tuttle R. 2000. Obituary: Sherwood Larned Washburn. *American Anthropologist* 102: 865–69.

Washburn SL. 1942. Skeletal proportions of adult langurs and macaques. *Human Biology* 14: 444–72.

Washburn SL. 1963. Behavior and human evolution. In *Classification and Human Evolution*, ed. SL Washburn, pp. 190–203. Chicago: Aldine Publishing Co.

Washburn SL. 1967. Behaviour and the origin of man. *Proceedings of the Royal Anthropological Institute of Great Britain and Ireland*: 21–27.

Washburn SL. 1968. One hundred years of biological anthropology. In *One Hundred Years of Anthropology*, ed. JO Brew, pp. 97–115. Cambridge, MA: Harvard University Press.

Washburn SL. 1983. Evolution of a teacher. *Annual Review of Anthropology* 12: 1–24.

Washburn SL, DeVore I. 1961. The social life of baboons. *Scientific American* 204: 62–71.

Yerkes RM, Yerkes AW. 1929. *The Great Apes*. New Haven, CT: Yale University Press.

4

PRIMATOLOGY IN ANTHROPOGENIC CONTEXTS

An Extended Evolutionary Approach

Although primatology first had a home in American anthropology, in the 1970s and 1980s the primary source of theoretical and methodological insight shifted to the biological sciences, whereby evolutionary approaches to the study of behavior and standardized, quantitative observation techniques came to dominate. In current times, however, field primatologists are finding themselves in ecological contexts that are heavily influenced by *human* primates; a reality that has been perceived by some to be a conundrum for studies using evolutionary approaches to behavior. How can one accurately study behavior in "unnatural" environments? Indeed, humans may drive environmental changes at a faster rate and across larger spatial scales than non-anthropogenic sources of environmental variation (Sih 2013), but this does not mean that understanding how animals respond to such change should be outside of our realm of interest, and for that matter, our expertise. Accordingly, my goal in this chapter is to show how and why it is important to examine primate behavior in anthropogenic settings. To do so, we can still invoke evolutionary approaches, and thereby, elucidate key processes central to biological anthropology (e.g., adaptation). But, primatology can also benefit from greater engagement with components of an extended evolutionary approach, such as phenotypic plasticity and niche construction. Among evolutionary biologists there have been recent conversations that their field's conceptual framework – the Modern Synthesis[1] – is "in need of significant extension" (Pigliucci 2007: 2743).[2] The same may be true for primatology.

Primatology's Transformation

As a result of Hooton's vision and Washburn's efforts, a stream of anthropologically trained primatologists headed to the field in the 1960s, amassing more and more information on the lives of primates in their native habitats. And, within these expanding datasets, primatologists began to see that there was tremendous variability in primate behavior both between and within species. For example, as we saw in Chapter 3, Thelma Rowell (1966) challenged the notion of species-typical behavior with her observations of olive baboons (*Papio anubis*) living in forested environments in Uganda: compared to their savannah-dwelling conspecifics studied by Washburn and DeVore, forest baboons exhibited smaller daily path lengths and home ranges, spent less time travelling, and existed in more "relaxed and friendly groupings" (p. 363). The most logical next step then was to attempt to explain *why* such variability exists – and to do so, primatologists looked beyond anthropology, to ecology and evolutionary biology, for guidance. The surrounding environment in which primates lived, for example, might explain the patterns of behavior that were emerging. And, notably, some of the early anthropological primatology works (e.g., DeVore & Washburn 1963) began to hint at such interconnections.

Further insight for contextualizing observed primate behavior in its ecological setting was provided by the work of ethologists John Crook and Steve Gartlan (1966) and John Eisenberg and colleagues (1972) on "grades of social organization" that linked habitat type, diet, and diurnal activity with group size, group composition, and mating patterns among primates. Inherent to this approach was an emphasis on the functional significance of behavior, and hence, attention to the selection pressures that account for the evolution of primate social organization and behavior. Building upon Crook and Gartlan's work, zoologist Thomas Struhsaker (1969) expanded our understanding of the link between ecology and behavior beyond the savannah by studying the group composition, group structure, social roles, social communication of rain forest-dwelling cercopithecines in Cameroon.

Interest in linking behavior and ecology continued to expand among primatologists in the 1970s and 1980s, particularly as the fields of sociobiology and behavioral ecology gained prominence. Sociobiology – a gene centric paradigm championed by E.O. Wilson in his 1975 "Sociobiology: The New Synthesis" – sought to explain social behavior in terms of self-maximization; that is, individuals engage in behavior that enables them to more successfully pass on their genes. Although this framework grew out of ethological studies on insects, birds, and other non-primates, the idea was that it could be used to study all aspects of primate social behavior – from kinship patterns, competition and cooperation, to inter- and intrasexual relationships (Fedigan & Strum 1999). Behavioral ecology, sociobiology's cousin of sorts, which became conceptually unified as a field with John Krebs and Nicholas Davies' (1978) publication of the 1st edition of "Behavioral Ecology: An Evolutionary Approach" (Owens 2006), was also concerned with the adaptive or survival value of behavior, but aimed to do so by contextualizing behavior in relation to ecological and social conditions. In other words, like sociobiology, behavioral ecology sees the individual as the unit of selection, but it also holds that ecological (and social) factors constrain an individual's attempt to maximize fitness.[3] Drawing from these fields, evolutionary theory was invoked to explain the evolution of primate sociality and the variation we see in primate behavior via two major evolutionary forces, natural selection and sexual selection, with the ecological environment as the backdrop. And, with this, *primate* behavioral ecology was born.

Accompanying the adoption of evolutionary approaches to the study of primate behavior was a methodological shift away from rich description toward systematic techniques that aimed to minimize observer bias, specified sampling regimes, and prioritized quantitative data and analysis. Jeanne Altmann's (1974) classic paper, "Observational Study of Behavior: Sampling Methods," for example, outlined a series of methods (e.g., focal-animal sampling, scan sampling, etc.) that could be used to systematically and quantitatively record the behavior of animals in the "real world," and thereby enable researchers to achieve "reliable and valid results" (Altmann 1984: 51).[4] This shift to quantitatively-oriented techniques was not surprising given that the central objective of evolutionary approaches to behavior is to identify, quantify, and model fitness costs and benefits. In addition, as primatologists began to accumulate more and more data across the primate order and in varied field settings, standardized techniques were needed to ensure proper comparative analysis to explain primate behavioral variation. Moreover, given the primacy of the ecological environment to these approaches, primatologists also needed to turn to ecology for guidance on how to assess the key features of these environments that influence primate behavior (e.g., forest structure and composition).

And therein lies the transformation: anthropological primatology answered Washburn's call in his "New Physical Anthropology" to incorporate evolutionary theory as an explanatory tool, and in doing so, primatology strengthened its theoretical and methodological affinities outside of anthropology. The result was a transformed primatology: anthropologically inspired, biologically framed and conducted.[5] In the 1980s, primate behavioral ecology became the reigning paradigm for this transformed primatology, and this paradigm along with its evolutionary offshoots have maintained their reign ever since.

The Reigning Paradigm: Primate Behavioral Ecology

While primate behavioral ecology uses an evolutionary perspective to explore primates' responses to their environment more broadly, its offshoot, primate socioecology, seeks to explain variation in primate social relationships in relation to ecology. The socioecological model was developed as a way to predict how important ecological factors, such as the distribution of resources and predation risk, shape primate social structure, primate social organization, and the competitive regimes within and between groups.[6] Because female fitness is constrained by access to food due to greater energetic investment in reproduction (Trivers 1972), female relationships are the primary focus of socioecological studies. On the other hand, because male fitness is constrained by access to mates, their social strategies are believed to be a product of sexual selection. While the model was initially developed to account for interspecific variation in female social relationships, within-species variation in behavioral patterns is now more widely appreciated (Chapman & Rothman 2009; Strier 2009). Collectively primate behavioral ecology and primate socioecology represent the primary paradigm used by behavioral primatologists today.

Since the synthetic socioecological model first emerged, some studies have found a good fit between the model's predictions and observed behavior.[7] For example, there is support for a relationship between the intensity of competition and group size and the concomitant effects on other behavioral parameters such as home range size (Chapman et al. 1995). Other studies, however, have found mismatches. For instance, despite high population density experienced by Sulawesi moor macaques, females are not involved in between-group competition as is predicted by the model (Okamoto & Matsumura 2002). These mismatches have led some scholars to question the model's predictive power and argue for alternative explanations for observed behavior, such as demographic variability (Struhsaker 2008) and phylogenetic relatedness (Thierry 2008; Di Fiore & Rendall 1994). While phylogeny cannot tell us why a behavioral pattern evolved in the first place, it can explain why we see behavioral similarities in some primate lineages despite the fact that the species live in varying ecological conditions. For example, phylogeny can explain why we see female philopatry and kin relations across the cercopithecine subfamily and how the patterning of social style of *Macaca* spp. is linked to phyletic lineage (Ménard 2004; Thierry 2007). Others have called for attempts to improve the model's heuristic value and explanatory power by addressing some of the methodological issues of its application, such as actually quantifying key variables crucial to the model (e.g., spatial and temporal patterns of food availability) rather than relying on general categorizations linking diet to competitive regimes (Koenig & Borries 2009; Clutton Brock & Janson 2012). Scholars have also suggested expanding the model's scope to consider other important ecological and social factors, such as infanticide and female counterstrategies (van Schaik & Janson 2000); disease (Huffman & Chapman 2009); increased attention to male social strategies and the potential role that males may play in altering modes of competition (Fashing 2001); and cognitive abilities (Cunningham & Janson 2007).

Another important variable that affects a primate's socioecology is anthropogenic disturbance. But, how and where do humans fit into the model? Indeed, the synthetic socioecological model does not explicitly incorporate the human dimension, and its influence on primate lives.[8] Instead, humans and their activities are often viewed as interferences. I call this the "unnatural problem," which I see as stemming from a couple of issues. The first issue concerns: what counts as "natural environments" and what is humans' place in nature? Since its inception, field primatology has been fundamentally concerned with the behavior and ecology of nonhuman primates in their natural environments. What is generally implied by "natural environments" is an ecological context that is free of human influence and a place where primates exhibit their "natural behavior" undisturbed from the human touch. But, what a number of the early researchers began to realize was that many of the populations they set out to study were not free of human influence. During his 1955 trip to Africa, Washburn himself was taken with the tame group of baboons who frequented his hotel in Victoria Falls, and how they had adapted to the circumstances: "you could see that they there making all of these sensible and learned adjustments" (DeVore & Washburn 1992: 420). Nonetheless, this stipulation – studying primates in "natural nature" – has predominately held even as field primatology sought theoretical and methodological insight from biology – a field that has generally situated humans as part of nature in an ecological sense and that explicitly recognizes humans as a major evolutionary force (e.g., Palumbi 2001). The second issue is about clarifying the level of analysis, proximate or ultimate, when using the behavioral ecological framework. Or, to avoid some of the ambiguity in the meaning of "ultimate," we need to clarify what question we are asking about behavior.[9] Here I am referring to the pioneering work of ethologist Niko Tinbergen (1963) who argued that there are four types of questions we can ask about behavior: its immediate causation, its development, its survival value, and its evolution.[10] Let's look at these issues in more detail in an effort to move beyond the "unnatural problem."

Beyond the "Unnatural Problem"

The "unnatural problem" is in part due to the view that human processes, such as the alteration of primate habitats, are interferences that result in evolutionary unstable situations, thereby rendering the socioecological model a poor fit for observed primate behavior.[11] This perspective stems from the notion that significant anthropogenic activity in zones of human-nonhuman primate sympatry is recent on an evolutionary time scale, and hence, of marginal concern in models that seek to explain the evolution of social behavior in primates. The resulting perspective is that these study populations and their "unnatural" anthropogenic environments pose an epistemological and methodological dilemma for field primatology.[12] Washburn, for example, although enamored with the tame baboon troops, chose to focus on those living in more natural conditions. No humans allowed! Except for human observers, of course – but this is another issue.

One of the issues with this perspective, however, is that it ignores the fact that the archaeological record points to considerable antiquity of human (hominin)- primate interaction. For example, there is accumulating evidence of hominin predation on nonhuman primates across the world. Based on cut-mark evidence on remains of *Paleopropithecus* bones from Taoloambiby, Madagascar, paleoanthropologists Laurie Godfrey and William Jungers (2003) argue that human hunting likely played a role in the extinction of at least some of the large-bodied subfossil lemur species. The concentration and distinctive breakage pattern of bones from a now extinct gelada monkey species (*Theropithecus oswaldi*) found at an Acheulian hand axe site in Kenya suggest they were butchered by *Homo erectus* (Shipman et al. 1981). In Venezuela, zooarchaeological remains of howler monkeys found in association with stone tools from at least 3,000 years ago point to the possibility that howlers were prey for pre-Colombian peoples (Urbani 2005). In all of these examples, there is "sufficient time depth for the co-evolution of predator-prey dynamics" between hominins and other primates (Sponsel 1997: 153). If I could talk to the early pioneers of primatology today, they would probably agree, for many of them recognized humans as predators and noted that humans had been so for a long time.[13] Take, for example, the common (mis)perception that the fleeing response of primates upon encountering human observers represents "wild" behavior – Washburn and Hamburg (1968) quite reflexively, given their desire to study natural behavior, recognized that this response was actually a marker of human impact, namely, the result of human predation. Today, primates continue to be key subjects of prey in contemporary areas of human-nonhuman primate sympatry (Shepard 2002; Urbani 2005), but humans are typically viewed as external actors – "hunters" or "poachers" – rather than a natural component of the faunal community of particular ecosystems, in other words, as predators themselves (Sponsel 1997).[14] This difference in perspective is important because it shapes how we think about and do research. For example, with regard to the socioecological model, the former perspective accounts for why humans are seen as only playing a role in eliminating natural predators (cf. Sterck 1998) and generally excluded from primate socioecological analyses of predator-prey dynamics.[15]

Other forms of human-primate symbiosis also show considerable antiquity, and hence, ecological and evolutionary importance. Drawing from diverse sources of evidence – fossil, archaeological, genetic, and behavioral, biological anthropologist Stephanie Spehar and colleagues (2018) demonstrate how humans have influenced the distribution, population size, and ecology of orangutans (*Pongo* spp.) for at least 70,000 years. In Lopé, Gabon, it is believed that three hominoid genera (*Pan*, *Gorilla*, and *Homo*) coexisted for a period of at least 60,000 years, possibly competing for similarly favored plant foods (Tutin & Oslisly 1995). Wheatley et al. (1996) point out the fact that new habitats (e.g., riverine secondary forests) were created for long-tailed macaques (*Macaca fascicularis*) in Southeast Asia as result of the widespread practice of swidden agriculture over thousands of years. In Bali, Indonesia, the long-term anthropogenic alteration of the landscape has shaped the population structure (by affecting dispersal patterns, and fragmenting and homogenizing populations) and the transmission of pathogens (with large, heavily provisioned populations unexpectedly showing a lessened prevalence and intensity of gastrointestinal parasites) in the resident populations of long-tailed macaques (Lane et al. 2011; Lane de-Graaf et al. 2014). In the Yucatan, México, it has been hypothesized that the current distribution of spider monkeys around archaeological ruins may be related to pre-Colombian peoples' modification of ecological landscapes; namely, the Maya cultivated tree species in close proximity to their settlements that in turn were preferred foods for spider monkeys (see Cormier & Urbani 2008).

The potential longevity of these examples of sympatry therefore challenges us to reconsider whether present forms of human-primate symbiosis necessarily represent an *unnatural* situation (Southwick et al. 1965).[16] Moreover, in the contemporary era it is becoming increasingly difficult, if not impossible, to find a primate population that has not experienced some form of anthropogenic influence, whether it be directly through predation or foraging encounters or indirectly through anthropogenic habitat alteration or destruction. For example, ecologist Serge Wich and colleagues (2012) estimated that 54 percent of the distribution of Bornean orangutans (*Pongo pygmaeus*) lies in areas dominated by human activities (e.g., logging concessions and oil palm and tree plantations). In the Neotropics, the primary distribution of the pied tamarin (*Saguinus bicolor*) lies in the urban areas of Manaus, Brazil (Farias et al. 2015). Other examples of contemporary human-primate sympatry abound, and yet, open the existing general textbooks on primate behavior and ecology, and you will find little to no mention of humans save a paragraph or two on threats to primate conservation. These texts have made humans as co-inhabitants and co-participants in primate ecologies invisible.[17]

The second component to the "unnatural problem" concerns the purpose of the synthetic socioecological model. While some aim to elucidate the evolution of behavior, the behavioral ecological framework, from which the model derives, also enables us to ask questions about the adaptive significance of behavior. At this level of analysis the anthropogenic effect is not a dilemma but rather an opportunity to explore primate adaptability at work and how anthropogenic processes may be important socioecological variables in their own right. Indeed, "the burden of proof falls on those researchers who disbelieve that the same selective mechanisms operate on both 'wild' and food-enhanced populations" (Wheatley et al. 1996: 203). By embracing the study of primate behavior in human-modified environments, we are joining other evolutionary scientists, including behavioral ecologists, who see such environments as opportunities to explore *contemporary evolution* and increase our knowledge of the adaptive potential of our study species (e.g., Schroeder et al. 2011; Sih 2013; Wong & Candolin 2015).[18] It is also worth pointing out that anthropogenic settings – in which rapid environmental change is occurring – should be considered methodologically robust for they are in essence serving as natural experiments for the study of ecological and evolutionary processes (Strier 2009).[19] For field primatology, for which an experimental research design can be challenging, this reality should be a welcomed silver lining. By moving beyond "the unnatural problem," we are rejecting the contention that our ability to investigate and interpret the adaptive significance of behavior is compromised by the anthropogenic touch (cf. Caro & Sherman 2011). As I note above though, to do so requires clarity regarding the level of analysis we are investigating and adopting a dynamic view of adaptation.

Adaptation: Adaptability

Adaptation is a central concept in the evolutionary sciences, and yet, it is variably defined, often in conflicting ways. While many contend that genetic changes and the action of natural selection are required for adaptation, others adhere to a view of adaptation that is genotype- and history-free.[20] For example, zoologists Tim Clutton-Brock and Paul Harvey's (1979) definition of adaptation – "a difference between phenotypic traits which increases inclusive fitness" (p. 114) does not require that the differences have a genetic basis; that is, they define adaptation "by its effects rather than its causes" (p. 115). Biologists Hudson Reeve and Paul Sherman (1993: 9) offer a similar definition of adaptation – "a phenotypic variant that results in the highest fitness among a specified set of variants in a given environment." This definition recognizes that a phenotype can be produced in other ways than by genetic means alone and makes no references to how the phenotype is produced.[21] This is not to say that we cannot appreciate that "contemporary populations are each the products of their own particular ecological histories" (Henzi & Barrett 2005: 223); but rather that we make no claims about the specific historical processes or evolutionary mechanisms that produced the patterns we see today. Such a definition may be particularly appropriate for behavioral differences when we cannot be sure of a genetic origin (Clutton-Brock & Harvey 1979).

By adopting this definition, recent behaviors in novel environments (e.g., behavioral responses by primates living in human-modified habitats) are not "problems" for our field; rather, they are considered adaptations as long as they result in higher fitness compared to behavioral alternatives (Reeve & Sherman 1993). Contemporary environments in which primates live that include humans and the effects of their behavior (e.g., habitat fragmentation, agroforestry, urbanized landscapes, etc.) may differ from the environments in which the primates evolved in terms of the specific environmental details, but not necessarily in terms of the fundamental trade-offs they present (Smith et al. 2000). Moreover, we expect that primates have the ability to recognize and appropriately respond to those trade-offs. Shirley Strum noted this early on when demonstrating how community ecology factors, both human and non-human (i.e., human-induced landscape transformations and competition from ungulates) explain the variation in fecundity she observed among female olive baboons over the course of a 10-year period: "baboons don't distinguish between the sources (natural versus human) of environmental change in their responses to change" (Strum 2012: 4).[22] Thus, in anthropogenic contexts, such as the forest-farm matrix or urban spaces, we can assess the ecological conditions in terms of the tradeoffs they present, and ask, under these conditions, "how would a well-adapted individual behave?" (Altmann & Altmann 2003: 414). For example, zoologist Gaelle Fehlmann and colleagues (2017) used this line of reasoning to develop predictions about they how they expected human-induced environmental changes to affect male baboon (*Papio ursinus*) foraging strategies in the Cape Peninsula, South Africa. At this site, urban spaces provide high-energy food resources, yet accessing these resources can be risky due to the presence of field managers who are tasked with deterring "raiding" baboons. They found that male baboons negotiate the costs and benefits of their human-modified habitat by spending more of their time at the urban edge where they engage in low-activity behaviors, and then engage in short, high-activity and risky forays into urban areas to access high-energy resources (e.g., trash bins) when they can.

Primatologist Anindya Sinha and colleagues (2005) used a similar approach to explain the emergence of a type of social structure not expected for a seasonally breeding cercopithecine primate such as the bonnet macaque, *Macaca radiata*: one-male, multi-female groups. The key factors involved in this example include the presence and behavior of humans and behavioral flexibility on part of the bonnet macaques in the Bandipur-Mudumalai wildlife sanctuaries in southern India. While all three populations of bonnet macaques under study experienced some level of provisioning, two of them lived in close proximity to human habitation, while the third only received food from tourists as they passed along the road. Accordingly, because the food from tourists was high quality, but unpredictable and clumped in distribution due to seasonally varying levels of tourism in the sanctuaries, access to it led to high levels of competition among females eventually resulting in group fission. Smaller groups of females emerged from the fissioning that could then be monopolized by a single adult male. And, voilà, unimale groups emerge! In contrast, the groups from the populations living in proximity to human habitation that experienced more regular access to human food were not under pressure to fission, and hence, maintained a social structure more typical of the species (Sinha et al. 2005).

To summarize, I support a contemporary primatology that moves beyond the "unnatural problem" and considers anthropogenic settings an important and worthy context in which to examine primate behavior and ecology. Such an approach adopts a dynamic view of adaptation, recognizes humans as major selective agents, and seeks to explain behavioral variation in contemporary environments. Importantly, by studying primates in human-modified environments, we are not just asking proximate questions about the mechanisms behind the behavior that we see (i.e., immediate causation); we are asking questions of a hybrid nature encompassing at least two of Tinbergen's "four questions" and reflective of a more pluralistic approach. We are still concerned with adaptive significance, but the context is not adaptation to past environments, but adaptability in present ones. In other words, this approach shifts the focus to seeing adaptability unfold in the now, that is, the ongoing evolutionary process. I say "at least two" because the ontogeny or development of flexible behavioral responses to anthropogenic impacts is equally critical to our understanding of behavioral plasticity and its evolutionary and ecological consequences. Accepting anthropogenic settings as worthy study contexts also paves the way for a contemporary primatology guided by an "extended evolutionary approach" that orients its attention to additional processes that drive evolution (e.g., behavioral plasticity and niche construction). We will explore these elements in more detail below by looking at some examples that support such a shift in approach.

Behavioral Plasticity in the Spotlight

As we have seen, over the years primatology came to recognize that variation in behavior is not limited to differences between species; there is also a considerable amount of intraspecific variation that exists and warrants our investigation and explanation. Primates living in anthropogenic landscapes are no exception to this. Phenotypic plasticity represents a powerful and appropriate framework to explain intraspecific variation in primate behavior and explore the adaptive significance of behavior (Ghalambor et al. 2010), particularly in the context of human-induced environmental change. At its simplest, phenotypic plasticity refers to an individual's ability to change their phenotype in response to differing environmental conditions. These transformations can involve life history, morphological, and physiological traits, but by far, behavior is viewed as among the "most labile of phenotypes" (Foster & Sih 2013:1003). The study of phenotypic plasticity is by no means a new endeavor; it has been a key focus for many evolutionary biologists for decades, if not longer, who have provided evidence that a wide diversity of organisms – from the orange jewelweed to fruit flies – express plastic responses to their environments (Pigliucci 2001). Along with behavioral ecologists more generally, behavioral primatologists are increasingly seeing the value of examining behavior through this framework.[23]

Behavioral plasticity is distinct from constructs that primatologists are familiar with, such as "generalists" versus "specialists," as these classifications tend convey a static view of response, and hence, overlook important sources of flexibility such as social learning and innovation (Wright et al. 2010).[24] Instead, by using a behavioral plasticity lens, we recognize the active role that primates play in negotiating environments that are rarely fixed. We also see that, following Sih (2013) there are likely multiple stages of response to human-altered environments; there will be the initial plastic response, followed by learning that allows them to persist until evolutionary responses (i.e., genetically-produced adaptations) are possible. With this approach, a key goal is to explain the variation that we see in response to anthropogenic change: why do some individuals respond well, while others don't? Natural selection is expected to favor flexibility when the fitness benefits associated with such flexible responses outweigh the costs (Dingemanse & Wolfe 2013). So, for primates in anthropogenic settings, we can make predictions about what we expect to see based on, for example, life history theory and socioecological factors (e.g., temporal and spatial distribution of food and resulting competitive regimes), and then determine if we see differences in behavior vis-à-vis anthropogenic factors (e.g., access to provisioned foods or crop lands). If we do, we can then ask: how do these differences affect survival and reproductive rates?[25] While some plastic responses may be adaptive, others may not be, particularly if human-induced environmental change results in a mismatch between the cues primates use to assess habitat quality (e.g., spatially concentrated food in a plantation) and the actual quality of the habitat (i.e., high rates of retaliation by humans guarding the plantation) or if benefits accrued come at the expense of other fitness-related activities, such as foraging and reproduction (Wong & Candolin 2015).

A new situation that has recently developed at my field site in Bantimurung-Bulusaraung National Park, South Sulawesi, Indonesia serves as a useful heuristic of how one might employ this framework. One of our moor macaque (*Macaca maura*) study groups has been habituated to human presence for many years due to occasional provisioning by park staff and regular following by researchers, but it has never been observed interacting with people traveling on the provincial road that intersects the park. Recently, though, the park management relocated a monitoring post to a location along the road within the home range of the group. Although this post is meant for park staff only, motorists often use it as a rest stop where they eat food and leave behind waste. This group is now spending a large proportion of its day along the road waiting for people in passing cars to throw them food and foraging in trash pits that collect along the roadside (Figure 4.1).

FIGURE 4.1 Moor macaques (*Macaca maura*) eating provisioned food from people in passing vehicles in Bantimurung-Bulusaraung National Park, Sulawesi, Indonesia. Note also the trash alongside the road. Photo credit: Kate Jameson.

Rather than viewing this as aberrant behavior, we can recognize it as behavioral plasticity – the emergence of a new behavioral tradition – and ask: Do we see individual variation in this behavior, and why? Personality, for example, may constrain how flexible individuals can be. We would also ask whether the costs associated with this behavior (e.g., being hit by cars and motorcycles, potential illness from human foods) are outweighed by the benefits (i.e., access to high energy foods with minimal energy expenditure) and what the evolutionary consequences of such flexibility are, at both individual and population levels. Focusing our attention to the flexibility in responses that primates make as they navigate increasingly anthropogenic spaces brings us closer to an alternative viewpoint that a number of primatologists have hinted at for some time: that we should be thinking more in terms of the "potential for modification" (Kummer 1971), that is, examining and explaining the *ranges* of behavior patterns that we see today under different circumstances (Strum & Fedigan 1999; Strier 2009). Not coincidentally, this perspective advocates for attention to a key component of the phenotypic plasticity framework: the behavioral reaction norm, which views behavior in terms of potential ranges rather than as categorical entities (Ghalambor et al. 2010).

Another research trend in behavioral plasticity that could be incorporated into an extended evolutionary approach for primatology is the role of behavioral plasticity in driving evolutionary change. If behavioral plasticity enables an organism to extend its range to include other habitats, then it is possible that patterns of selection can also be altered (Ghalambor et al. 2010). For example, using an experimental design, biologist Jonathan Losos and colleagues (2004) determined that *Anolis* lizards (*Anolis sagrei*) on islands in the Bahamas became more arboreal over time following the introduction of a terrestrial predator, the curly-tailed lizard (*Leiocephalus carinatus*). However, because some of the *Anolis* lizards still remained terrestrial and succumbed to predation, selection favored attributes that made the lizards more difficult to catch and consume: longer legs in the case of males, and larger bodies for females. The authors predict that if there is greater shift toward arboreality in future generations, selection will, in turn, also shift to favor those with attributes better fit for arboreal substrates (in this case, shorter limbs). For primates, anthropogenic settings may provide the experimental context in which to explore the role of behavioral plasticity in such ecological and evolutionary processes; however, we will need data across much longer time frames to see shifting patterns and sources of selection. Building more long-term study sites will make these kinds of investigations possible moving forward.

Human, Primate, and Human-Primate Niche Construction

In 2004, biologists Robert Sapolsky and Lisa Share shared some fascinating primate news with the scientific community: a different social style had emerged within a baboon group following an outbreak of bovine tuberculosis (TB) that killed the highest ranked and highly aggressive males from the group, leaving behind a cohort of males that were less aggressive. The outbreak of TB originated from infected meat that had been dumped in an open garbage pit close to a tourist lodge where some of the males, in particular the most aggressive, began to frequent. The new social style that emerged was characterized by high rates of grooming and affiliation with females and a more relaxed dominance hierarchy among males – a pattern that was maintained for (at least) a decade following the deaths of the aggressive males (Sapolsky & Share 2004). The authors also determined that the males who were left behind had lower glucocorticoid levels, a finding that suggests a less stressful environment following the outbreak and subsequent deaths of the high-ranked males. Not only does this example aptly demonstrate how human activities (i.e., the construction of a tourist facility with open garbage pits) can have long-lasting effects on both baboon social behavior and biology, it also illustrates the outcomes of another phenomenon of ecological and evolutionary significance that is relevant to an extended evolutionary approach for primatology: niche construction.

In ecological terms, a niche refers to the ecological space an organism occupies and the role, function, and impact of the organism in the environment.[26] With species coexistence in mind, ecologist G. Evelyn Hutchinson (1957) envisioned the niche as the multidimensional hyperspace of environmental factors affecting an organism, and further delineated the concept into the *fundamental niche* – the set of ecological conditions that enable an organism to survive and reproduce – and the *realized niche*, as what is left of the fundamental niche, so to speak, after considering the possible ecological interactions a species has with other organisms in their shared environment (e.g., competition, parasitism, predation, etc.). *Niche construction* is the process by which organisms select and modify their niches in ways that have ecological and evolutionary consequences. Niche construction is viewed as not just a product of natural selection – a perspective touted by skeptics of an extended evolutionary approach – but as "an evolutionary process in its own right" (Day et al. 2003: 81).[27] The idea here is that through niche construction, organisms not only shape their environment but also the selection pressures that act on them, their descendants, and other species (Day et al. 2003; Odling-Smee et al. 2003). Natural selection and niche construction are therefore recognized as co-directors in the evolution of organisms, with the former operating directly on genetic variation and the latter operating indirectly by shaping the selective environments (Kendal et al. 2011).

Niche construction at work has been documented across a diverse array of taxa – from ant colonies whose movements affect the flow of nutrients in ecosystems, marine plants that provide protected habitat for other species (e.g., kelp forests), to our own genus, *Homo*. For example, it has been argued that various modes of human subsistence constitute forms of niche construction (Rowley-Conwy & Layton 2011). Hunter-gatherer groups, such as the Nukak of the Colombian Amazon, who consume wild fruits at camp sites – that is, away from where they grow – in turn, create new areas of intensified fruit tree production, thereby altering a key selective pressure for other animals: food availability. Additionally, the burning of vegetation for cultivation can increase the productivity of certain foods as well as provide new or preferred habitat for other animals, whose population numbers could in turn increase. In Western Australia, for example, although hill kangaroo (*Macropus robustus*) experience predation pressure from Aborginal hunters, they also benefit from anthropogenic burning practices that result in concentrated patches of emerging shoots (Codding et al. 2014). Niche construction is also evident via hunting which can lead to the removal of competitor species (e.g., decimation of large carnivore populations in Europe following the arrival of modern humans), "niche deterioration" as a result of overhunting, and "niche enhancement" resulting from human-induced introductions of prey species to non-native habitats (Rowley-Conwy & Layton 2011: 852).

An extended evolutionary approach for primatology that incorporates the niche construction perspective firmly positions humans within the ecologies of primates, and enables us to consider how human niche construction may be altering the selection pressures affecting primates. Examples abound beyond the TB case in baboons described above! Biologist Tom Waite and colleagues (2007), for example, found that unlike their counterparts living in a protected area, Hanuman langurs (*Semnopithecus entellus*) in the urban center of Jodhpur, India were buffered from the impact of a severe drought by human provisioning. Urbanization also comes with increased noise pollution, which can act as a new selection pressure for urban dwelling wildlife. In Minas Gerais, Brazil, for example, researchers determined that a group of black-tufted marmosets (*Callithrix penicillata*) living in an urban park more frequently occupied quieter areas of the park, even when that meant forgoing areas with a greater abundance of food trees (Duarte et al. 2011). Indeed, as urbanization expands worldwide, there is increasing evidence of human-driven contemporary evolution in a diverse array of taxa, as well as evidence that phenotypic change is greater in urban environments compared to non-urban anthropogenic environments (Alberti et al. 2017). Using the niche construction framework, contemporary primatology is primed to investigate human-driven phenotypic and genetic change in primates. For example, in the case of the urban marmosets dealing with anthropogenic noise, future work could determine whether this anthropogenic pressure causes stress (measurable via a variety of biomarkers), and if so, whether there is evidence of adaptive divergence in behavior, physiology, or morphology among this urban dwelling group compared to other black-tufted marmoset populations. Another next step will be to move beyond using niche construction as a heuristic tool to describe patterns post-hoc and work toward testing predictions that derive from an extended evolutionary approach (see Laland et al. 2015).

An extended evolutionary approach for primatology also means attributing "a more central role of organisms in the evolutionary process" (Laland et al. 2015: 8), and hence, recognizing primates themselves as niche constructors. For example, Fuentes (2017) argues for greater attention to be paid to how primate social networks function as a mechanism of niche construction. Hypersociality and group living are key components of what it means to be a primate. Therefore, the selective forces impacting primates derive not only from the ecological environment but also the complex social ecologies in which they live. Importantly, primates are actors in these social landscapes, developing, learning, manipulating, and remembering the diverse array of social relationships they build. The niche construction framework can help us better understand how primates build their social networks and are impacted by them (Fuentes 2017).[28]

Finally, the niche construction perspective is a useful approach for examining the human-primate interface in anthropogenic spaces. In Bossou, Guinea, for example, anthropologist Kimberly Hockings and colleagues (2017) documented the dispersal of an important cash crop, cacao (*Theobroma cacao*), by a chimpanzee community across its home range, and importantly, beyond plantation boundaries (Figure 4.2). The behavioral flexibility of chimpanzees, in this case, flexible feeding on cultivated foods, is creating new pockets of anthropogenically-sourced, primate-planted resources, which in turn may be impacting the distribution and abundance of other tropical plants, and the other wildlife that feed on them (see also Sengupta et al. 2015). Moreover, the possibility that humans may benefit from this flexible feeding strategy and the resulting niche construction points to the importance of looking not just at humans as niche constructors or primates as niche constructors in anthropogenic spaces, but humans and primates as *co*-niche

FIGURE 4.2 An adult male chimpanzee (*Pan troglodytes verus*) feeding on ripe cacao fruit at Bossou, Guinea. Photo credit: Nicola Bryson-Morrison.

constructors in their overlapping niches (cf. Fuentes 2010). To explore this co-construction necessitates an extended evolutionary approach like the one we have discussed above; but it also requires one that incorporates both human and non-human primate perspectives and positions the "human-primate community" as its unit of investigation. That approach is *ethnoprimatology*. As we will see in the next chapter, with the human-primate interface as its central focus, ethnoprimatology embraces the reality that in many areas of the world primates in anthropogenic spaces is the "new natural." And, importantly, this reality is not the end point; but rather, the starting point for cutting-edge, mixed-methods, intra- and interdisciplinary research. Follow me; let's check it out!

Notes

1 The Modern Synthesis, at its most simplest, refers to the integration of Charles Darwin and Alfred Wallace's theory of evolution by natural selection with Gregor Mendel's laws of inheritance. Since this paradigm was first coined in the 1940s, it has incorporated the work of other notable evolutionary biologists (e.g., Ernst Mayr and Thomas Dobhansky) and emergent fields (e.g., genetics).

2 I should note that there is disagreement among evolutionary biologists on whether the key elements of an "extended evolutionary" approach – i.e., niche construction and phenotypic plasticity – are actually missing from more "standard evolutionary approaches" (see Point and Counterpoint commentary by Laland et al. 2014 and Wray et al. 2014). Regardless, these frameworks are quite certainly underutilized in behavioral primatology, and hence represent, at least for a primatology, an extended primatological approach.

3 This alone represented a major shift given the group-centric view inherent to the structural-functionalist approach of early anthropological primatology (see Gilmore 1981; Strum & Fedigan 1999) and to the primatology practiced by scholars from other disciplines. Robert Hall (1968), an early field primatologist trained in psychology, saw the group as the primary unit of study: "the success of baboons…is a consequence, not of individual prowess, but of the fitness of the social organization."

4 Altmann (1974) used this term, "real world," to signify behavior observed in field settings in contrast to laboratory research. See Montgomery (2005, 2015) for further use of this term.

5 This transformation is, of course, linked to the shift from using primates as referential models toward the application of key principles of evolutionary theory to explain primate behavioral evolution more broadly (see Tooby & DeVore 1987).

6 For the socioecological model see Wrangham (1980), van Schaik (1989), Isbell (1991), and Sterck et al. (1997). Definitions of primate social structure and primate social organization have been reversed over the years, and still vary depending on which scholar you reference. Here I follow Rowell (1972) and define "social structure" as group composition and the spatial patterning of individuals via dispersal and "social organization" as the social interactions exchanged among individuals and the social relationships that ensue.

7 For the "synthetic socioecological model," I am following Thierry (2008): Box 1 page 94.

8 Anthropologist Tracie McKinney (2015) devised a classification system to describe the different levels and types of anthropogenic disturbance that could be incorporated into behavioral ecological/socioecological models.
9 The ambiguity stems from the oft-conflated current function with the reason it evolved in the first place. See Thierry (2005), Laland et al. (2013), and Bateson & Laland (2013) for further discussion of the drawbacks associated with Mayr's dichotomous model of proximate-ultimate causation.
10 Survival value is also referred to as the function, the adaptive significance, or the "current utility" (see Bateson & Laland 2013) of behavior. Also, I should note that, technically speaking, Tinbergen contributed a fourth problem, "ontogeny" (development) to the other three already proposed by Julian Huxley (1942).
11 See van Schaik (1996).
12 Early sociocultural anthropologists experienced a similar "dilemma" when they realized that their supposedly isolated study (human) populations were not isolated after all (Riley 2013).
13 See Hall (1968), Washburn & Hamburg (1968).
14 For example, in his paper on the importance of primates to the Matsingenka's subsistence and worldview, anthropologist Glenn Shepard (2002: 125) critiques John Terborgh's contradictory view in *Requiem for Nature* on what should and should not be a part of nature: "Given Terborgh's assertion that top predators 'are an essential component of healthy ecosystems' (1990, p.57), it seems ironic that he called for the removal of indigenous people, top predators *par excellence*, from parks."
15 Other fields in the social and natural sciences are keenly aware of the extreme impact human predation can have on animal populations, both terrestrial and marine. A case in point is Chris Darimont and colleagues' (2015) piece in *Science* titled "The unique ecology of human predators." By analyzing marine exploitation terrestrial mammal rates across all continents (except for Antarctica), the authors determined that humans hunt at significantly higher rates than other predators, with a distinct focus on adult prey. This pattern identifies humans not only as key predators, but more aptly as "super predators."
16 For example, Charles Southwick and colleagues (1965: 158) wrote: "It is, in fact, difficult to assess what the natural habitat of the rhesus [monkey] actually is. It has lived in close ecological contact with man for centuries, and in frequency and persistence this commensal relationship in villages, towns, temples, and roadsides represents a natural relationship."
17 Here I am referring to *Primates in Nature* (Richard, 1985), *The Life of Primates* (Nystrom & Ashmore 2008), and *Primate Behavioral Ecology* (Strier 2011). However, as I mention in Chapter 1, Strier's 5th edition of *Primate Behavioral Ecology* (2017a) is the first general textbook to situate humans as part of the ecology of primates.
18 Contemporary evolution refers to evolutionary change that occurs within a relatively short time frame – that is, on an ecological time-scale, or less than a few hundred generations. Because of the magnitude of anthropogenic impact in contemporary time, human activities are generally associated with contemporary evolution (see Stockwell et al. 2003; Carroll et al. 2007).
19 In fact, one could argue that "humans may transform the 'natural' order by reshaping the very disposition and biology of animals" (Arluke & Sanders 1996: 18).
20 See Reeve and Sherman (1993:10)

21 In their review of the transformation of behavioral field studies, Stuart Altmann and Jeanne Altmann (2003) adopt a similar definition of adaptation.

22 It is also worth pointing out that in their 1982 paper, Shirley Strum and David Western called for the inclusion of humans in primate community ecology – "the need to consider primate ecology within a broader community framework which includes human activity" (p. 61) – a perspective also espoused by the ethnoprimatological approach.

23 See Ghalambor et al. (2010), for behavioral ecology, and Henzi and Barrett (2005); Chapman and Rothman (2009); Nowak and Lee (2013); Kamilar and Baden (2014); Kappeler and Fitchel (2015); Strier (2017b) for primatology.

24 Some prefer to distinguish between 'behavioral plasticity' and 'behavioral flexibility' in terms of the time course over which changes occur, with the latter referring to short-term environmental effects (see Piersma & Drent 2003; Rubenstein 2016). Here, for simplicity's sake, I label any behavioral transformation, regardless of time frame, as behavioral plasticity.

25 I review an illustrative case study on brown bears in Turkey by ecologist Gabriele Cozzi and colleagues (2016) in Chapter 7.

26 Following animal ecologist, Charles S. Elton (1927).

27 But see Wallach (2016) and Gupta et al. (2017) who challenge the explanatory power of the niche construction framework and its proponents' claims that niche construction is an evolutionary process in its own right.

28 See also Flack (2006), MacKinnon and Fuentes (2011), and Kurvers et al. (2014).

References

Alberti M, Correa C, Marzluff JM, Hendry AP, Palkovacs EP, et al. 2017. Global urban signatures of phenotypic change in animal and plant populations. *Proceedings of the National Academy of Sciences* 114: 8951–56.

Altmann J. 1974. Observational study of behavior: sampling methods. *Behavior* 49: 227–67.

Altmann J. 1984. Observational sampling methods for insect behavioral ecology. *Florida Entomologist*: 50–56.

Altmann SA, Altmann J. 2003. The transformation of behaviour field studies. *Animal Behaviour* 65: 413–23.

Arluke A, Sanders CR. 1996. *Regarding Animals*. Philadelphia: Temple University Press.

Bateson P, Laland KN. 2013. Tinbergen's four questions: an appreciation and an update. *Trends in Ecology & Evolution* 28: 712–18.

Caro T, Sherman PW. 2011. Endangered species and a threatened discipline: behavioural ecology. *Trends in Ecology & Evolution* 26: 111–18.

Carroll SP, Hendry AP, Reznick DN, Fox CW. 2007. Evolution on ecological time-scales. *Functional Ecology* 21: 387–93.

Chapman CA, Rothman JM. 2009. Within-species differences in primate social structure: evolution of plasticity and phylogenetic constraints. *Primates* 50: 12–22.

Chapman CA, Wrangham RW, Chapman L. 1995. Ecological constraints on group size: An analysis of spider monkey and chimpanzee subgroups. *Behavioral Ecology and Sociobiology* 36: 59–70.

Clutton-Brock T, Harvey P. 1979. Comparison and adaptation. *Proc. R. Soc. Lond. B* 205: 547–65.

Clutton-Brock T, Janson C. 2012. Primate socioecology at the crossroads: past, present, and future. *Evolutionary Anthropology: Issues, News, and Reviews* 21: 136–50.

Codding BF, Bird RB, Kauhanen PG, Bird DW. 2014. Conservation or co-evolution? Intermediate levels of aboriginal burning and hunting have positive effects on kangaroo populations in Western Australia. *Human Ecology* 42: 659–69.

Cormier LA, Urbani B. 2008. The ethnoprimatology of spider monkeys (*Ateles* spp.): From past to present. In *Spider Monkeys: Behavior, Ecology, and Evolution of the Genus Ateles*, ed. CJ Campbell, pp. 377–403. Cambridge: Cambridge University Press.

Cozzi G, Chynoweth M, Kusak J, Çoban E, Çoban A, et al. 2016. Anthropogenic food resources foster the coexistence of distinct life history strategies: year-round sedentary and migratory brown bears. *Journal of Zoology* 300: 142–50.

Crook JH, Gartlan JS. 1966. Evolution of primate societies. *Nature* 210: 1200–03.

Cunningham E, Janson C. 2007. A socioecological perspective on primate cognition, past and present. *Animal Cognition* 10: 273–81.

Darimont CT, Fox CH, Bryan HM, Reimchen TE. 2015. The unique ecology of human predators. *Science* 349: 858–60.

Day RL, Laland KN, Odling-Smee FJ. 2003. Rethinking adaptation: The niche-construction perspective. *Perspectives in Biology and Medicine* 46: 80–95.

De Vore I, Washburn S. 1992. An interview with Sherwood Washburn. *Current Anthropology* 33: 411–23.

DeVore I, Washburn SL. 1963. Baboon ecology and human evolution. In *African Ecology and Human Evolution*, ed. FC Howell, F Bourliere, pp. 335–67. Chicago: Aldine.

Di Fiore A, Rendall D. 1994. Evolution of social organization: A reappraisal for primates by using phylogenetic methods. *Proceedings of the National Academy of Sciences of the United States of America* 91: 9941–45.

Dingemanse NJ, Wolf M. 2013. Between-individual differences in behavioural plasticity within populations: causes and consequences. *Animal Behaviour* 85: 1031–39.

Duarte MH, Vecci MA, Hirsch A, Young RJ. 2011. Noisy human neighbours affect where urban monkeys live. *Biology Letters* 7: 840–42.

Eisenberg JF, Muckenhirn NA, Rudran R. 1972. The relation between ecology and social structure in primates. *Science* 176: 863–74.

Elton CS. 1927. *Animal Ecology*. Chicago: University of Chicago Press.

Farias IP, Santos WG, Gordo M, Hrbek T. 2015. Effects of forest fragmentation on genetic diversity of the Critically Endangered primate, the pied tamarin (*Saguinus bicolor*): Implications for conservation. *Journal of Heredity* 106: 512–21.

Fashing PJ. 2001. Male and female strategies during intergroup encounters in guerezas (*Colobus guereza*): Evidence for resource defense mediated through males and a comparison with other primates. *Behavioral Ecology and Sociobiology* 50: 219–30.

Fehlmann G, O'Riain MJ, Kerr-Smith C, Hailes S, Luckman A, et al. 2017. Extreme behavioural shifts by baboons exploiting risky, resource-rich, human-modified environments. *Scientific Reports* 7: 15057.

Flack JC, Girvan M, De Waal FB, Krakauer DC. 2006. Policing stabilizes construction of social niches in primates. *Nature* 439: 426–29.

Foster SA, Sih A. 2013. Behavioural plasticity and evolution. *Animal Behaviour* 85: 1003.

Fuentes A. 2010. NaturalCultural encounters in Bali: Monkeys, temples, tourists, and ethnoprimatology. *Cultural Anthropology* 25: 600–24.

Fuentes A. 2017. Niche construction. In *The International Encyclopedia of Primatology*: John Wiley & Sons, Inc.

Ghalambor CK, Angeloni LM, Carroll SP. 2010. Behavior as phenotypic plasticity. In *Evolutionary Behavioral Ecology*, ed. DF Westneat, CW Fox, pp. 90–107. Oxford: Oxford University Press.

Gilmore H. 1981. From Radcliffe-Brown to sociobiology: Some aspects of the rise of primatology within physical anthropology. *American Journal of Physical Anthropology* 56: 387–92.

Godfrey LR, Jungers WL. 2003. The extinct sloth lemurs of Madagascar. *Evolutionary Anthropology* 12: 252–63.

Gupta M, Prasad N, Dey S, Joshi A, Vidya T. 2017. Niche construction in evolutionary theory: the construction of an academic niche? *Journal of genetics* 96: 491–504.

Hall KRL. 1968. Social organization of the Old-World monkeys and apes. In *Primates: Studies in Adaptation and Variability*, ed. PC Jay, pp. 7–30. New York: Holt, Rinehart, and Winston.

Henzi SP, Barrett L. 2005. The historical socioecology of savanna baboons (*Papio hamadrayas*). *J. Zool., London* 265: 215–26.

Hockings KJ, Yamakoshi G, Matsuzawa T. 2017. Dispersal of a human-cultivated crop by wild chimpanzees (*Pan troglodytes verus*) in a forest–farm matrix. *International Journal of Primatology* 38: 172–93.

Huffman MA, Chapman CA, eds. 2009. *Primate Parasite Ecology: The Dynamics and Study of Host-Parasite Relationships*. Cambridge: Cambridge University Press.

Hutchinson GE. 1957. Concluding remarks. *Cold Spring Harbor Symposium Quantitative Biology* 22: 415–27.

Huxley J. 1942. *Evolution: The Modern Synthesis*. London: George Allen & Unwin.

Isbell LA. 1991. Contest and scramble competition: patterns of female aggression and ranging behavior among primates. *Behavioral Ecology* 2: 143–55.

Kamilar JM, Baden AL. 2014. What drives flexibility in primate social organization? *Behavioral Ecology and Sociobiology* 68: 1677–92.

Kappeler PM, Fichtel C. 2015. Eco-evo-devo of the lemur syndrome: did adaptive behavioral plasticity get canalized in a large primate radiation? *Frontiers in zoology* 12: S15.

Kendal J, Tehrani JJ, Odling-Smee FJ. 2011. Human niche construction in interdisciplinary focus. *Philosophical Transactions of the Royal Society B* 366: 785–92.

Koenig A, Borries C. 2009. The lost dream of ecological determinism: Time to say goodbye?…or a white queen's proposal? *Evolutionary Anthropology* 18: 166–74.

Kummer H. 1971. *Primate Societies: Group Techniques of Ecological Adaptation*. Chicago: Aldine Publishing Company.

Kurvers RH, Krause J, Croft DP, Wilson AD, Wolf M. 2014. The evolutionary and ecological consequences of animal social networks: emerging issues. *Trends in Ecology & Evolution* 29: 326–35.

Krebs JR, Davies NB. 1978. *Behavioral Ecology: An Evolutionary Approach*. Oxford: Blackwell Scientific Publications.

Laland K, Wray GA, Hoekstra HE. 2014. Does evolutionary theory need a rethink? Point: Yes, urgently. *Nature* 514: 161–64.

Laland KN, Odling-Smee J, Hoppitt W, Uller T. 2013. More on how and why: Cause and effect in biology revisited. *Biology & Philosophy* 28: 719–45.

Laland KN, Uller T, Feldman MW, Sterelny K, Müller GB, et al. 2015. The extended evolutionary synthesis: Its structure, assumptions and predictions. *Proc. R. Soc. B* 282: 20151019.

Lane KE, Holley C, Hollocher H, Fuentes A. 2011. The anthropogenic environment lessens the intensity and prevalence of gastrointestinal parasites in Balinese long-tailed macaques (*Macaca fascicularis*). *Primates* 52: 117–28.

Lane-deGraaf K, Fuentes A, Hollocher H. 2014. Landscape genetics reveal fine-scale boundaries in island populations of Indonesian long-tailed macaques. *Landscape ecology* 29: 1505–19.

Losos JB, Schoener TW, Spiller DA. 2004. Predator-induced behaviour shifts and natural selection in field-experimental lizard populations. *Nature* 432: 505–08.

MacKinnon KC, Fuentes A. 2011. Primates, niche construction, and social complexity: The roles of social cooperation and altruism. In *Origins of altruism and cooperation*, pp. 121–43: Springer.

McKinney T. 2015. A classification system for describing anthropogenic influence on non-human primate populations. *American Journal of Primatology* 77: 715–26.

Ménard N. 2004. Do ecological factors explain variation in social organization? In *Macaque Societies*, ed. B Thierry, M Singh, W Kaumanns, pp. 237–62. Cambridge: Cambridge University Press.

Montgomery GM. 2005. Place, practice and primatology: Clarence Ray Carpenter, primate communication and the development of field methodology. *Journal of the History of Biology* 38: 495–533.

Montgomery GM. 2015. *Primates in the Real World*. Charlottesville: University of Virginia Press.

Nowak K, Lee PC. 2013. "Specialist" primates can be flexible in response to habitat alteration. In *Primates in Fragments: Complexity and Resilience*, ed. LK Marsh, CA Chapman, pp. 199–211. New York: Springer.

Nystrom P, Ashmore P. 2008. *The Life of Primates*. Upper Saddle River, NJ: Pearson Prentice Hall.

Odling-Smee FJ, Laland KN, Feldman MW. 2003. *Niche Construction: The Neglected Process in Evolution*. Princeton, NJ: Princeton University Press.

Okamoto K, Matsumura S. 2002. Intergroup encounters in wild moor macaques. *Primates* 43: 119–25.

Owens IP. 2006. Where is behavioural ecology going? *Trends in Ecology & Evolution* 21: 356–61.

Palumbi SR. 2001. Humans as the world's greatest evolutionary force. *Science* 293: 1786–90.

Piersma T, Drent J. 2003. Phenotypic flexibility and the evolution of organismal design. *Trends in Ecology & Evolution* 18: 228–33.

Pigliucci M. 2007. Do we need an extended evolutionary synthesis? *Evolution* 61: 2743–49.

Pigliucci M. 2001. *Phenotypic Plasticity: Beyond Nature and Nurture*. Johns Hopkins University Press: Baltimore.

Reeve HK, Sherman PW. 1993. Adaptation and the goals of evolutionary research. *The Quarterly Review of Biology* 68: 1–32.

Richard AF. 1985. *Primates in Nature*. New York: WH Freeman and Co.

Riley EP. 2013. Contemporary primatology in anthropology: Beyond the epistemological abyss. *American Anthropologist* 115: 411–22.

Rowell TE. 1966. Forest living baboons in Uganda. *Journal of Zoology* 149: 344–64.

Rowell TE. 1972. *Social Behaviour of Monkeys*. Baltimore: Penguin Books.

Rowley-Conwy P, Layton R. 2011. Foraging and farming as niche construction: Stable and unstable adaptations. *Philosophical Transactions of the Royal Society B: Biological Sciences* 366: 849–62.

Rubenstein DI. 2016. Anthropogenic impacts on behavior: The pros and cons of plasticity. In *Conservation Behavior: Applying Behavioral Ecology to Wildlife Conservation and Management*, ed. O Bergel-Tal, D Saltz, pp. 121–46. Cambridge: Cambridge University Press.

Sapolsky RM, Share LJ. 2004. A pacific culture among wild baboons: Its emergence and transmission. *PLoS Biology* 2: e106.
Schroeder J, Nakagawa S, Hinsch M. 2011. Behavioural ecology is not an endangered discipline. *Trends in Ecology & Evolution* 26: 320–21.
Sengupta A, McConkey KR, Radhakrishna S. 2015. Primates, provisioning and plants: Impacts of human cultural behaviours on primate ecological functions. *PloS one* 10: e0140961.
Shepard GH. 2002. Primates in Matsingenka Subsistence and World View. In *Primates Face to Face: Conservation Implications of Human-Nonhuman Interconnections*, ed. A Fuentes, L Wolfe, pp. 101–36. Cambridge: Cambridge University Press.
Shipman P, Bosler W, Davis KL, Behrensmeyer AK, Dunbar R, et al. 1981. Butchering of giant geladas at an Acheulian site [and Comments and Reply]. *Current Anthropology* 22: 257–68.
Sih A. 2013. Understanding variation in behavioural responses to human-induced rapid environmental change: a conceptual overview. *Animal Behaviour* 85: 1077–88.
Sinha A, Mukhopadhyay K, Datta-Roy A, Ram S. 2005. Ecology proposes, behavior disposes: Ecological variability in social organization and male behavioral strategies among wild bonnet macaques. *Current Science* 89: 1166–78.
Smith EA, Borgerhoff Mulder M, Hill K. 2000. Evolutionary analyses of human behaviour: a commentary on Daly & Wilson. *Animal Behaviour* 60: F21–F26.
Southwick CH, Beg MA, Siddiqi MR. 1965. Rhesus monkeys in North India. In *Primate Behavior: Field Studies of Monkeys and Apes*, ed. I DeVore, pp. 111–59. New York: Holt, Rinehart, & Winston.
Spehar SN, Sheil D, Harrison T, Louys J, Ancrenaz M, et al. 2018. Orangutans venture out of the rainforest and into the Anthropocene. *Science Advances* 4: e1701422.
Sponsel LE. 1997. The human niche in Amazonia: Explorations in ethnoprimatology. In *New World Primates: Ecology, Evolution, and Behavior*, ed. WG Kinnzey, pp. 143–65. New York: Aldine Gruyter.
Sterck EHM. 1998. Female dispersal, social organization, and infanticide in langurs: Are they linked to human disturbance? *American Journal of Primatology* 44: 235–54.
Sterck EHM, Watts DP, van Schaik CP. 1997. The evolution of female social relationships in nonhuman primates. *Behavioral Ecology and Sociobiology* 41: 291–309.
Stockwell CA, Hendry AP, Kinnison MT. 2003. Contemporary evolution meets conservation biology. *Trends in Ecology & Evolution* 18: 94–101.
Strier KB. 2009. Seeing the forest through the seeds: Mechanisms of primate behavioral diversity from individuals to populations and beyond. *Current Anthropology* 50: 213–28.
Strier KB. 2011. *Primate Behavioral Ecology, 4th edition*. New York: Pearson.
Strier KB. 2017a. *Primate Behavioral Ecology, 5th Edition*. New York: Routledge.
Strier KB. 2017b. What does variation in primate behavior mean? *American Journal of Physical Anthropology* 162: 4–14.
Struhsaker TT. 1969. Correlates of ecology and social organization among African cercopithecines. *Folia Primatologica* 11: 80–118.
Struhsaker TT. 2008. Demographic variability in monkeys: Implications for theory and conservation. *International Journal of Primatology* 29: 19–34.
Strum SC. 2012. Darwin's monkey: Why baboons can't become human. *American Journal of Physical Anthropology* 149: 3–23.
Strum SC, Fedigan L. 1999. Theory, method, gender, and culture: What changed our views of primate society? In *The New Physical Anthropology: Science, Humanism, and Critical*

Reflection, ed. SC Strum, DG Lindburg, D Hamburg, pp. 67–105. Upper Saddle River, NJ: Prentice Hall.

Strum SC, Western JD. 1982. Variations in fecundity with age and environment in olive baboons (*Papio anubis*). *American Journal of Primatology* 3: 61–76.

Thierry B. 2005. Integrating proximate and ultimate causation: Just one more go! *Current Science* 89: 1180–83.

Thierry B. 2007. Unity in diversity: Lessons from macaque societies. *Evolutionary Anthropology* 16: 224–38.

Thierry B. 2008. Primate socioecology, the lost dream of ecological determinism. *Evolutionary Anthropology* 17: 93–96.

Tinbergen N. 1963. On aims and methods of ethology. *Ethology* 20: 410–33.

Tooby J, DeVore I. 1987. The reconstruction of hominid behavioral evolution through strategic modeling. In *The Evolution of Human Behavior: Primate Models*, ed. WG Kinzey, pp. 183–237. New York: State University of New York Press.

Trivers RL. 1972. Parental investment and sexual selection. In *Sexual Selection and the Descent of Man*, ed. B Campbell, pp. 136–79. London: Heinemann.

Tutin CEG, Oslisly R. 1995. *Homo*, *Pan* and *Gorilla*: Co-existence over 60,000 years at Lope in Central Gabon. *Journal of Human Evolution* 28: 597–602.

Urbani B. 2005. The targeted monkey: A re-evaluation of predation on New World primates. *Journal of Anthropological Sciences* 83: 89–109.

van Schaik CP. 1989. The ecology of social relationships amongst female primates. In *Comparative Socioecology*, ed. V Standen, RA Foley, pp. 195–218. Oxford: Blackwell Scientific Publications.

van Schaik CP. 1996. Social evolution in primates: The role of ecological factors and male behaviour. *Proceedings of the British Academy* 88: 9–31.

van Schaik CP, Janson CH, eds. 2000. *Infanticide by Males and its Implications*. Cambridge: Cambridge University Press.

Waite TA, Chhangani AK, Campbell LG, Rajpurohit LS, Mohnot SM. 2007. Sanctuary in the city: Urban monkeys buffered against catastrophic die-off during ENSO-related drought. *EcoHealth* 4: 278–86.

Wallach E. 2016. Niche construction theory as an explanatory framework for human phenomena. *Synthese* 193: 2595–618.

Washburn SL, Hamburg DA. 1968. Aggressive behavior in Old World monkeys and apes. In *Primates: Studies in Adaptation and Variability*, ed. PC Jay, pp. 458–78. New York: Holt, Rinehardt & Winston.

Wheatley BP, Harya Putra DK, Gonder MK. 1996. A comparison of wild and food-enhanced long-tailed macaques (*Macaca fascicularis*). In *Evolution and Ecology of Macaque Societies*, ed. JE Fa, DG Lindburg, pp. 182–206. Cambridge: Cambridge University Press.

Wich SA, Gaveau D, Abram N, Ancrenaz M, Baccini A, et al. 2012. Understanding the impacts of land-use policies on a threatened species: Is there a future for the Bornean Orang-utan? *PLOS ONE* 7: e49142.

Wilson EO. 1975. *Sociobiology, The Modern Synthesis*. Cambridge, MA: Belknap.

Wong B, Candolin U. 2015. Behavioral responses to changing environments. *Behavioral Ecology* 26: 665–73.

Wrangham RW. 1980. An ecological model of female-bonded primate groups. *Behaviour* 75: 262–99.

Wray GA, Hoekstra HE, Futuyma DJ, Lenski RE, Mackay TF, et al. 2014. Does evolutionary theory need a rethink? Counterpoint: No, all is well. *Nature* 514: 161–64.

Wright TF, Eberhard J, Hobson E, Avery ML, Russello M. 2010. Behavioral flexibility and species invasions: the adaptive flexibility hypothesis. *Ethology Ecology & Evolution* 22: 393–404.

5

PRIMATOLOGY IN ANTHROPOGENIC CONTEXTS

The Emergence of Ethnoprimatology

One of my favorite primate fieldwork stories is one told by a colleague of mine who works in the Neotropics. Like many primatologists trained in biological anthropology, they had set out to do the bread and butter of traditional field primatology: examine primate behavioral ecology in natural, pristine environments. Their site was ideal by many measures; the forest was lush and primate dense, and it was located hundreds of miles from human occupation. Perfection. But, not long after their arrival, they came upon something that made them rethink just how pristine this forest was: archaeological evidence – from stone axe heads and pottery sherds to tool sharpening sites – revealing that humans, at one point in time, had made a living in the area, and hence, likely played an important role in shaping the rainforest that exists today.[1] It was this experience that led them to appreciate the ethnoprimatological approach. Simply put, ethnoprimatology is the study of the multifaceted ways the histories, ecologies, lives, and livelihoods of humans and primates intersect. Ethnoprimatology recognizes the reality and potential longevity of human-impacted environments and views them as worthy and interesting contexts in which to examine primate behavior and ecology (Fuentes & Hockings 2010).

The term "ethnoprimatology" was first coined by Leslie Sponsel (1997), an ecological and sociocultural anthropologist, who envisioned a fruitful space where primate ecologists and human ecologists could find mutual interest. Drawing largely from the field of ecological anthropology,[2] which during the 1990s was diversifying, Sponsel (1997) outlined six areas in which the human-primate interface could be examined: the comparative ecology of primates and humans; predation ecology – human as primate predators; symbiotic ecology – other symbiotic relationships between humans and primates; cultural ecology – the cultural relevance of primates; ethnoecology – traditional knowledge about primates;

and conservation ecology – the conservation implications of the human-primate interface.

Around the same time, Bruce Wheatley, a four-field trained anthropologist, began researching long-tailed macaques (*Macaca fascicularis*) in Bali, Indonesia, and realized he could not fully understand the ecology of these macaques without considering the human cultural context. In his book, *The Sacred Monkeys of Bali*, Wheatley (1999) demonstrates the mutual relevance of primatology and sociocultural anthropology by exploring the important role long-tailed macaques play in Balinese culture, the longevity and prevalence of the human-macaque interface in Bali, and the implications of the human-macaque relationships for conservation. He offered the term "cultural primatology" as a way to unite primatology and sociocultural anthropology in the study of the interactions between humans and primates. Eventually, *ethnoprimatology* emerged as the preferred term, presumably to distinguish the field from the study of cultural traditions among nonhuman primates (cf. McGrew 2007).

At the same time, I think it is important to clarify how the use of the "ethno" prefix for ethnoprimatology differs compared to the standard anthropological usage. In sociocultural anthropology, the prefix *ethno-* (e.g., *ethnoscience*, *ethnobotany*, *ethnobiology*) serves to distinguish traditional, indigenous ways of knowing from that of Western scientific understandings. While ethnoprimatology can indeed encompass local knowledge and perceptions about primates, it is more than that; it also encompasses a much more literal meaning, with "ethno + primatology" signifying the intersection of human and nonhuman primate lives and livelihoods across both time and space (Riley 2018).

Although Sponsel's and Wheatley's emphases differed slightly – i.e., the former focused on more ecologically-oriented themes while the latter focused more on cultural elements, they both reconceptualized what is meant by "primates in nature."[3] They recognized humans as part of nature (rather than separate from), and viewed primate behaviors that reflect human influences not as aberrant but as integral and interesting parts of their ecological strategies.

The Ethnoprimatological Scope

During the 2000s, interest in the approach increased dramatically, and ethnoprimatology came to encompass Sponsel's more ecologically-focused themes and the more cultural elements of Wheatley's vision.[4] In addition to building upon these pioneering works, scholars also recognized the ethnoprimatological salience of other research areas that already had established, multidisciplinary trajectories and incorporated them into ethnoprimatology's purview. These research areas included: the interface of humans and other primates in tourist settings (e.g., provisioning; aggressive interactions; Fa 1991; Saj et al. 1999) and at the forest-farm interface (e.g., crop foraging; Strum 1994; Hill 2000), and human-human primate disease transmission (Daszak et al. 2001; Wolfe et al. 2005). The result of this amalgamation of foci was the emergence of a scope of inquiry that was based first and foremost in anthropology, but was simultaneously integrative and interdisciplinary in nature, incorporating theory and method from primate ecology, disease ecology, sociocultural anthropology, environmental anthropology, and conservation ecology. Let's look at ethnoprimatology's scope in more detail to explore how it differs from existing perspectives, contributes to key objectives of biological anthropology (albeit in new ways – a theme that continues from Chapter 4), but also expands the purpose of primatology, thereby giving it a home in *anthropology*, not just its biological quadrant.

Human-Primate Communities: Sympatry and Adaptability

Ethnoprimatology represents an ideal approach to address key principles and processes near and dear to the heart of biological anthropology: sympatry and adaptation during the course of primate and human evolution. As we learned in Chapter 4, it is becoming increasingly clear from the fossil record that sympatry (overlapping geographic ranges) characterized much of the primate and human evolutionary past (Ward 2003). Sympatry with humans also characterizes the contemporary state of many extant primate populations. Examining the ways that human and nonhuman primates coexist and negotiate overlap today not only sheds light on patterns of sympatry in the past, but also eludicates patterns of adaptability and flexibility. To do so, ethnoprimatology offers the "human-primate community" as the focal point and draws from a number of theoretical frameworks and perspectives to examine it. The conception of "human-nonhuman primate communities" and the various ecological relationships (e.g., competition, predation, commensalism, etc.) that characterize these communities stem from Sponsel's vision of integrated human and primate ecologies, but also from community ecology.[5]

In ecology, a "community" is defined as a group of species living in a specified place and time (Vellend 2010). The subfield of *community ecology* focuses on the interactions between species and their environment. Examining human-primate coexistence also relies on the standard theoretical framework used by primatologists to explain how ecological factors shape behavior: behavioral ecology. The primary difference is that humans and the ecological consequences of their actions are considered key components of a primate species' ecology, and hence, are figured into predictions about expected responses. With this scaffolding in place, ethnoprimatology is primed to ask the following questions regarding human-primate communities: Why are they sympatric? What is the nature of their ecological relationship (i.e., competition, predation, commensalism, mutualism, etc.)? How do they influence one another's behavior and ability to persist? How do they adapt to the ecological changes wrought by one another?

The sharing of space experienced by human-nonhuman primate communities can often translate into overlap in resource use. To understand how resource use patterns overlap and the implications of this overlap for future co-existence necessitates an integrated, mixed-methods approach. And, this is what ethnoprimatology provides. For example, in the Lindu highlands of Central Sulawesi, Indonesia, I assessed overlap in resource use by conducting both ethological observations of the feeding ecology of the Tonkean macaques (*Macaca tonkeana*) and ethnographic interviews with villagers on their knowledge and use of forest resources (Riley 2005). I determined that of the 113 tree species listed by villagers, 17 of them were utilized by both humans and the Tonkean macaques. I also dug deeper to explore the importance of the tree species for which there was overlap and found that five species that were among the top 10 most salient tree species listed by villagers constituted more than 80 percent of the Tonkean macaques' diet (Riley 2007).[6] And, one of those five resources stood out in particular: the Aren palm, *Arenga pinnata*. The Aren palm, also more widely known as the "sugar palm," was the most important food resource for one of my study groups; the fruits comprised 52 percent of their diet. From the human perspective, the palm is considered such an important resource because of the many benefits and products it provides. Hair fibers from the stalks are used to construct household items, the palm fronds are used for roofing of traditional huts, and the sap from the tree is tapped to produce palm sugar or various forms of alcoholic beverages (Figure 5.1). What I learned, however, is that many of these culturally salient products are easily obtained from the tree without its felling, meaning that the manner in which villagers exploit the palm ensures its persistence, and thus may *positively* impact the Tonkean macaque's ability to persist human-modified landscapes (Riley 2007). In some cases, though, the conservation implications of overlapping resource use are not so positive. In their study of overlapping resource use between Tanala village residents and the lemurs of Ranomafana National Park, Madagascar, primatologist Patricia Wright and colleagues (2005) found that of the 235 taxa of plants used by the Tanala, 56 of them were consumed by lemurs and human use of 60 percent of those 56 taxa involved felling and destruction of the plant, thereby inhibiting future use by lemurs. That being said, the authors acknowledge that if plant extraction for household use is maintained at traditional, low levels, then it will unlikely pose a threat to lemur survival.

Across the diverse array of settings where humans and primates interface, one ethnoprimatological lesson learned is that the standard definition of "resources" needs to be expanded (Riley 2018). "Resources" typically refers to wild foods found in primates' "natural habitat." However, in contemporary urban settings, primates may consume temple offerings, forage in human garbage dumps, and steal food from people's homes and shops. The foods they acquire are typically human foods and hence not "natural" per se, but nonetheless represent important sources of nutrition with concomitant implications for health, energetics, and population demography. For example, in Bangladesh, primatologists M. Firoj Jaman and Michael Huffman (2013) found that urban rhesus macaques spent less time feeding and more time resting than those living in rural areas, presumably

FIGURE 5.1 Lindu villager "Papa Sar" collecting palm wine (*tuak*, Bahasa Indonesia or *tule* in Bahasa Lindu) from his Aren tree (*Arenga pinnata*). The sap collects in bamboo, which are fixated at the terminal ends of cut flower trunks. Tomado, Lindu, Central Sulawesi, Indonesia. Photo credit: Erin P. Riley.

because the regular supply of provisioned foods they receive (e.g., chick peas, bread, biscuits, bananas) enable them to meet their energetic requirements more quickly. Another important difference was group size; the urban group was twice the size of the rural group, likely reflecting enhanced fecundity, but potentially leading to increased population density and competition with conspecifics as well as future conflict with urban human residents. Similarly, "resources" for macaque populations living in tourist settings in Bali, Indonesia, and Gibraltar are predominately provisioned foods; 70 percent of the long-tailed macaque population's diet in Padangtegal, Bali was provisioned food and in the Gibraltar Nature Reserve, provisions accounted for 76 percent of the Barbary macaque diets (Fuentes et al. 2007). The level and organized nature of provisioning at these sites have led to the occurrence of macaque obesity and extraordinary increases in population size (Fuentes et al. 2007; Lane et al. 2010) (Figure 5.2).

In rural settings, the primary form of ecological change occurring across the distribution of extant primate populations is the conversion of habitat to agricultural land. While the resulting matrix habitat may be unsuitable for some wildlife (Gascon et al. 1999), for many primate species, these habitat transformations result in the creation of new potential sources of food, and hence, new situations in

FIGURE 5.2 An obese adult long-tailed macaque (*Macaca fascicularis*) at the Uluwatu temple site, Bali, Indonesia. Photo credit: Erin P. Riley.

which overlapping resource use can occur.[7] Although there are some exceptions, generally, the crops planted are meant for human consumption, meaning they are highly palatable and have important macro- and micronutrients. Accordingly, you can bet they are equally fit for consumption by other primates. And, if we think about how long humans have been modifying landscapes to produce food (i.e., thousands of years), you can also bet that primates feeding on crops, at least in some places, likely has a deep history. Viewed through the lens that sees human influences as interferences that result in evolutionary unstable situations, crop feeding is an aberrant behavior. An alternative view, espoused by the ethnoprimatological approach, is that crop feeding exemplifies highly adaptive behavior, and hence, should be viewed as a foraging strategy in its own right (see Maples et al. 1976; Strum 1994). By taking this latter perspective, we are afforded an expanded understanding of primate feeding ecology – a key area of interest for primatology, and biological anthropology, more generally.

For primates that inhabit forest-agriculture ecotones, crop feeding may provide important high-quality foods during periods of wild food scarcity. For example, Zanzibar red colobus monkeys (*Procolobus kirkii*) were observed to consume more cultivated coconuts when there were fewer wild foods available (Siex &Struhsaker 1999). Shirley Strum (1994), one of the first anthropologists to examine crop feeding as an ecological strategy, determined that the most significant factor contributing to the development of crop feeding behavior in olive baboons (*Papio anubis*) was that there was less natural forage available to them. For chimpanzees (*Pan troglodytes verus*) in Boussou, Guinea, Kimberly Hockings and colleagues (2009) found that sugar fruit crops were foraged upon more frequently when forest fruit availability was reduced. What these studies highlight is the importance of crops to the primate's overall feeding ecology; namely, agricultural crops, particularly those available year-round (e.g., cacao), may serve as important fallback foods for primates living in habitats heavily impacted by anthropogenic pressures (McLennan 2013).[8]

In contrast to preferred foods, which are overselected relative to their abundance, fallback foods are typically abundant, available year-round, and hence, are exploited when preferred foods are not available (Marshall & Wrangham 2007). In recent years, biological anthropologists have been particularly interested in understanding the role fallback foods play in the ecology and evolution of primates. Ecologist Joanna Lambert and colleagues (2004), for example, proposed that the thick enamel of grey-cheeked mangabeys (*Lophocebus albigena*) has less to do with what they eat all the time, but rather, what they eat when their preferred foods are unavailable: food items that are more easily processed with thick enamel (e.g., hard bark and seeds). As noted by biological anthropologists Paul Constantino and Barth Wright (2009), Lambert and colleagues' analysis challenges us to think more critically about the selective forces driving primate feeding adaptations. With this in mind, in returning to the anthropogenic context in which many primates live, we see that if cultivated foods are becoming a new class of fallback foods, anthropogenic pressures too may be important to primate ecology and evolution.[9]

Viewing crop feeding – and for that matter, the acceptance of provisioned food – as foraging strategies also enables us to see how complex primate adaptive variability is. The alteration of primate habitats by human activities is generally thought to result in lower quality habitat characterized by a diminished food supply (Struhsaker 1997). While this can indeed be true in many cases, it may not *always* be the case, depending on the primate species in question, and more importantly, the nature and extent of the alteration. In other words, context matters. Anthropologist Matthew McLennan (2013) found, for example, that the chimpanzees living in human-disturbed environments in Bulindi, Uganda are not necessarily food-impoverished because agricultural crops (e.g., cacao) provide a year-round supply of food.

For food-enhanced primate populations, optimal foraging theory predicts that because of the clumped distribution and nutritional benefits of anthropogenic foods, primates are expected to show smaller home ranges, spend less time moving and feeding, and more time resting. Again, it may, however, not be as straightforward as that, as ecological responses can vary, both interspecifically and intraspecifically, depending on the anthropogenic context. Anthropologist Tracie McKinney (2011) found that food-enhanced white capuchins in Curú Wildife Refuge in Costa Rica had a larger home range and a similar activity budget to wild feeding conspecifics. Bruce Wheatley and colleagues (1996) found that food-enhanced groups of long-tailed macaques in Bali, Indonesia had smaller home ranges, spent more time resting, and less time moving compared to their wild conspecifics in Kalimantan. In Singapore, a food-enhanced long-tailed macaque group similarly spent less time resting, but more time moving and had a larger monthly home range compared to the "low anthropogenic" group (Sha & Hanya 2013). To understand these differences, we must dig deeper to understand the nature of the anthropogenic foods available to them. At the Monkey Forest site in Ubud, Bali, provisioning of long-tailed macaques is spatially concentrated; whereas in Singapore, provisioning is irregular, meaning that the macaques primarily obtain anthropogenic food from refuse sites that are spatially dispersed and where the availability of foods is unpredictable. These studies therefore demonstrate how critical it is for us to examine primate ecological responses across a range of anthropogenically-influenced habitats to better understand the extent of primate ecological flexibility and intraspecific variability and why they respond the way they do.

As I mentioned above, an interest in crop feeding is certainly not unique to ethnoprimatology, as there exists a rich body of work by social science scholars and conservation managers on crop feeding by wildlife and its impact on human livelihoods.[10] Nonetheless, because of its mixed methodological approach incorporating theoretical frameworks and methods from biological anthropology, environmental anthropology, and conservation ecology, ethnoprimatology is a particularly well-suited approach to study crop feeding and other ecological outcomes of human-primate sympatry. For indeed, anthropogenic habitat alteration can negatively impact primates' ability to survive, but primate responses to such pressures can, in turn, impose constraints on human livelihoods. The ethnoprimatological approach explores *both* the human and nonhuman primate perspective to understand patterns of primate ecological flexibility and how both perspectives intersect to affect human-primate coexistence. Therefore, this approach typically entails collecting and integrating quantitative data (e.g., crop loss measurements; the frequency and duration of raiding events; the relationship between crop feeding and wild food availability) and qualitative data (e.g., perceptions of and folklore about primates; why farmers act the way they do). Without measuring actual damage done by primates, a huge disparity may exist between reported and observed damage due to farmers overestimating the amount of crops loss. Moreover, people's perceptions of the extent of damage may be influenced by factors associated with the animals themselves (e.g., large body size, conspicuous behavior of most crop foraging primates). For example, in Lore Lindu National Park, while farmers believed the Tonkean macaque to be the most destructive crop forager, I found that forest rats were responsible for significantly more damage than macaques (Riley 2007).

Human-Primate Disease Ecology

Integrated human and primate ecologies not only translate into the sharing of space, but also the sharing of pathogens. Accordingly, the transmission of diseases between humans and other primates emerged as an important focus for ethnoprimatology. I should note that the study of human-primate disease transmission is certainly not unique to ethnoprimatology. There exists a robust, multidisciplinary research effort focused on host-parasite ecology, how the behavior and ecology of primates influences the spread of disease, and the links between anthropogenic disturbance and the emergence of infectious diseases across multiple scales.[11] An ethnoprimatological lens enables a detailed analysis of the specifics and nuances of the human-primate context in which pathogens emerge and spread. In other words, the ethnoprimatological approach digs deeper than the amorphous categories of "anthropogenic impact" or "anthropogenic factors" to explore what the specific contexts are, including who the people are and why they do what they do. This perspective is possible because of the anthropological base from which ethnoprimatology derives.

In Southeast Asia, pet ownership represents a salient context in which to examine human-primate disease ecology. For example, in Sulawesi, Indonesia, the ownership of pet monkeys is a relatively common practice among a number of ethnic and religious groups, whereby macaques (*Macaca* spp.) are usually acquired as infants by opportunistically trapping wild mother-infant pairs and/or from trapping used to minimize crop feeding (Jones-Engel et al. 2005). The captured infants are then typically fitted with a harness made from rope and tethered to a post or tree close to the owner's residence, where people have ample opportunity to interact with them (Figure 5.3). Biological anthropologist Lisa Jones-Engel and colleagues (2001), who have spearheaded ethnoprimatological research on human-primate disease transmission, have found evidence of exposure of both pet and wild Sulawesi macaques to endemic human pathogens, including measles, influenza A, and parainfluenza 1, 2, and 3: all respiratory viruses that are easily transmitted and highly contagious. It is believed that the pet macaques likely acquire the pathogens as a result of close interaction with their owners, and other members of the village. Moreover, the fact that the same human pathogens have been detected in wild populations suggests that the pet macaques may act as a sort of "vector" (from human to wild populations) since wild macaques are often attracted to pet ones (e.g., when females are in estrus) (Jones-Engel et al. 2005). Jones-Engel and colleagues (2004) have also determined that young pet macaques acquire intestinal fauna (e.g., enteric nematodes) more similar to human populations (with whom they have regularly contact) than wild populations of Sulawesi macaques.

Ethnoprimatologists have also examined disease transmission in temple and urban settings where humans and various primates regularly interface. In fact, religious temples are likely the most important setting for human-primate

FIGURE 5.3 Adult male olive baboon (*Papio anubis*) in a maize field eating a cob. Kekopey, Central Rift Valley of Kenya, Gilgil Baboon Project. Photo credit: Martin Western.

FIGURE 5.4 A pet juvenile booted macaque (*Macaca ochreata*) from Southeast Sulawesi, Indonesia. Note how he is tethered at his waist. Photo credit: Erin P. Riley.

pathogen exchange given that the number of people that come into contact with monkeys at these sites per year is in the millions (Fuentes et al. 2005). Throughout South and Southeast Asia, multiple species of primates (e.g., macaques, *Macaca* spp. and Hanuman langurs, *Semnopithecus entellus*) reside at religious temple sites, where they opportunistically acquire provisions from temple visitors via handouts or offerings, and in towns and densely populated cities where they may raid homes, scavenge for food in refuse areas, or benefit from provisioning from their fellow (human) city dwellers. In some urban locations, such as Indonesia's capital city of Jakarta, monkeys have very close contact with humans as a result of the role as street performers (Schillaci et al. 2005; Figure 5.5).[12] This context is distinct compared to pet ownership or temple settings; whereas in these settings the primates and their pathogens likely have a local origin, urban performing monkeys are typically acquired from large animal markets that "bring together a variety of species, primate and non-primate, from diverse geographic origins, each with its potentially unique burden of infectious agents" (Schillaci et al. 2005: 1306).

In these anthropogenic settings, it is generally assumed that closer and more frequent contact with humans results in a greater risk of interspecific disease transmission (Chapman et al. 2005), and indeed, there is evidence to support this assumption (e.g., anthropozootic transmission of gastrointestinal parasites to habituated gorillas; Graczyk et al. 2002). At the same time, it is also important to more deeply explore what "human contact" entails. With the human-primate interface being its focal point and its mixed-methods toolkit, the ethnoprimatological approach is primed to elucidate the ecological and anthropological details of "human contact." For example, in their research on long-tailed macaque populations across the island of Bali, Indonesia, biologist Kelly Lane and colleagues (2011) found a lessened prevalence and intensity of gastrointestinal parasites in those populations in close and regular contact with humans at tourist sites. They attribute this seemingly unexpected result to the fact that the nature of "human contact" at those sites entails heavy provisioning, which in turn may result in a reduced need to forage and hence less exposure to parasites (cf. Eley et al. 1989). An ethnoprimatological approach also enabled biological anthropologist James Loudon and colleagues (2006) to more fully appreciate the context for interspecies pathogen exchange in Beza Mahafaly Specical Reserve, Madagascar. At this site, they determined that differences in socioecology and behavior (e.g., group size and the extent of terrestriality) accounted for why ring-tailed lemurs (*Lemur catta*) had a higher prevalence of gastrointestinal parasites than sympatric Verraux's sifakas (*Propithecus verrauxi*). Because lemurs have more terrestrial behavior and have a generalist feeding strategy, they came into contact with and even consumed human feces when ranging through the camp area of the park. The lemurs were exposed to human feces because the local Mahafaly people use open-air latrines rather than pit-latrines as local tradition prohibits placing human waste into the ground. It was by making the *human-primate community* the unit

(a)

(b)

FIGURE 5.5 Long-tailed macaque (*Macaca fascicularis*) street performers, *topeng monyet*, in Manado, North Sulawesi (a) and in Tasik Jaya, Java (b) in Indonesia. Photo credit: (a) Lisa M. Danish, and (b) Little Fireface Project.

of investigation, which necessitates examining primate behavior *and* the human-cultural context, that this insight was possible.

In addition to revealing important details of the "anthropogenic" context in human-primate disease ecology, the ethnoprimatological approach demonstrates that humans act as selective pressures in their own right. Certainly, we know that disease and parasites can cause mortality in primates, thereby regulating populations, and have the potential to negatively affect reproductive status when, for example, parasites generate pathological effects and/or major energetic costs for their primate hosts (Chapman et al. 2005). For example, biological anthropologist Colin Chapman and colleagues (2006) found that parasite infection, in combination with nutritional stress, was linked to population decline in red colobus (*Piliocolobus tephrosceles*) living in forest fragments in Kibale National Park, Uganda. But what ethnoprimatological studies clearly show is how human bodies and behavior can shape the disease ecology of our closest living relatives, thereby having important proximate effects (e.g., their immediate health) and ultimate effects (e.g., humans affecting their evolutionary trajectories by shaping what diseases they are and are not susceptible to). And, of course, just as the transmission of disease is bidirectional, so too may be the evolutionary effects. For example, one reason suggested for the lack of reported cases of transmission of *Cercopithecine herpesvirus* 1 (Herpes B virus) from long-tailed macaques in Bali to people, despite frequent contact with Balinese temple workers and visitors, may be that temporal depth of their ecological sympatry has enabled the evolution of immunologic resistance in local (human) populations (Engel et al. 2002). Although this hypothesis remains untested, it is nonetheless intriguing.

Wait, Humans Matter Too? Cultural Facets of the Human-Primate Interface

Given that the ethnoprimatological approach was, at least partially, the brainchild of a sociocultural anthropologist, it is not surprising then that much of the early work in ethnoprimatology focused on the cultural dimension of the human-primate interface; that is, how primates figure into peoples' folklore, customs, and worldviews. And, there is good reason for such a focus; the cultural dimension of the human-primate interface is not without time depth. The cultural salience of primates in ancient Mesoamerica, for example, is well represented by the widely-known monkey figure of the Nazca Lines, a geoglyph complex believed to be constructed over 1,500 ya by the Nazca people in the arid coastal desert of southern Peru (Cormier & Urbani 2008). Reinhard (1988) reports that because monkeys are associated with places where water is available, the monkey figure (whose body configuration resembles a spider monkey, *Ateles* spp.) has been interpreted to either protect or symbolize water. Neotropical primates, in particular spider monkeys, are also prominently represented in pottery and stone figurines of Maya and Aztec material culture, respectively, from Mexico (Cormier & Urbani 2008).

Sociocultural anthropologists, in particular those interested in the ethnoecology and historical ecology of indigenous peoples of the Neotropics, are responsible for much of the early ethnoprimatological work on the cultural dimensions of contemporary human-primate interfaces. Motivated by an interest in examining "the relationship between different cultures and primate species" (Cormier & Urbani 2008: 378), these scholars, such as Loretta Cormier, Manuel Lizzaralde, and Glenn Shepard, elucidated how humans interact with primates in multiple domains, and determined that doing so is key to the communities' cultural identities. Primates have material importance, serving as important components to the diets of many indigenous Neotropical peoples. Dietary preferences, however, are not necessarily uniform; for instance, while the Guajá of Brazil prefer howler monkeys (Cormier 2002), the Bari of Venezuela and the Matsingenka of Peru, show preference for spider monkeys (Lizzaralde 2002; Shepard 2002). Here we see that an ethnoprimatological approach, in this case with attention to cultural factors that shape human behavior, helps explain this variability. Howlers are perceived as slow and lazy, and hence, the Bari and Matsingenka are fearful of the "contagion" of such characteristics (Cormier & Urbani 2008). Similarly, the Matsingenka regard capuchin monkeys to be mischievous and some believe that if children consume capuchin meat they will grow up to be dishonest (Shepard 2002). As Shepard (2002: 107) astutely notes, "understanding dietary preference and avoidance in any society involves an appreciation of empirical and ecological as well as cultural and symbolic aspects of food consumption;" it is this integrative perspective that the ethnoprimatological approach offers.

Another important symbolic theme prevalent in the Neotropics is that of "transformation" – whereby (nonhuman) primates derive from humans, or humans transform into other primates if cultural norms are not followed (Cormier & Urbani 2008). The mythologies of transformation, reviewed by anthropologists Loretta Cormier and Bernardo Urbani (2015), reveal how people perceive the natural world and humans' place it in it vis-à-vis other beings; they help us anthropologists better understand the observed patterns of human-primate interaction. Primates also play important roles in social domains.[13] For example, among the Guajá, orphaned monkeys (of all species) are incorporated into the social system whereby they are kept as pets and treated like dependent children. The social role of these young primates is multifarious. Because many of these young monkeys are nurtured by Guajá women, they serve to enhance fertility. They also play an important role in the socialization of children, and concomitantly, the reinforcement of gender roles; young girls practice carrying the infant monkeys, thereby simulating childcare, while young boys learn critical aspects of the monkey's behavior, such as vocalizations, that later enhance their hunting skills (Cormier 2002).

By the late 1990s/early 2000s, primatologists started paying attention to the value of the ethnographic lens in understanding primate populations and communities. As a budding primatologist in the late 1990s, I will never forget the first time I came across "ethnoprimatology" being presented at a conference. It was the 1999 American Society of Primatologists meeting in New Orleans where Loretta Cormier, a sociocultural anthropologist, gave a poster presentation on her work among the Guajá, titled "Cultural implications for Neotropical conservation on the Carú indigenous reserve, Maranhão, Brazil." This poster was the turning point for me. And I wasn't alone; Agustín Fuentes and Linda Wolfe, two biological anthropologists by training, took a chance by compiling much of this early work that I described above in their volume, *Primates Face to Face*, published in 2002.

Fast forward to the mid-to-late 2000s, and more and more primatologists began seeking training beyond the traditional boundaries of primatological practice. Coursework in ecological and environmental anthropology, research methods sociocultural anthropology, ecological economics, and conservation biology all provided insight for a new generation of ethnoprimatological projects that positioned the *human-primate community* as the unit of investigation. One guiding theoretical framework or one set of methods was no longer viewed as sufficient to examine the complexity of the human-primate interface. And, importantly, interest in the human dimension extended beyond traditional subjects of sociocultural anthropological inquiry (e.g., indigenous communities of the Neotropics) to include any and all human communities whose lives and livelihoods intersect with their primate neighbors. For example, in our respective work in Buton and Central Sulawesi, Sulawesi, Nancy Priston and I took a comprehensive approach to the study of the human-macaque interface in forest-farm matrix by observing macaque behavior, quantitatively measuring the extent of crop damage caused by macaques, and conducting ethnographic interviews with farmers and other villagers to access their perceptions of the interface and how and why they respond to it (Riley & Priston 2010). Biological anthropologist, Melissa Remis partnered with a sociocultural anthropologist, Rebecca Hardin, integrating ethnography of BaAka hunting camps with line transect, forest phenology, and gorilla socioecology data, to examine forest change and its impact on relationships between humans and primates in the Sangha Dense Forest Reserve, Central African Republic (Hardin & Remis 2006; Figure 5.6). Biological anthropologist Anna Nekaris and colleagues (2010) supplemented recorded international trade data with qualitative data garnered through interviews with individuals in source villages, trading posts, and wildlife markets to reveal the cultural factors driving the pet trade of *Loris* and *Nycticebus* in Sri Lanka, Cambodia, and Indonesia. More recently, anthropologist Sindhu Radhakrishna (2018) examined changes in the nature of human-primate relations as reflected in Tamil poetry from six periods in Tamil literary history, thereby adding time depth and a new methodological tool (i.e., literary analysis) for accessing human perceptions of primates to the ethnoprimatological approach.[14]

FIGURE 5.6 BaAka foragers of the Central African Republic prepare a blue duiker (*Philantomba monticola*), a small antelope, after net hunting. Duiker and monkey species are the primary prey species for hunters in this region. Photo credit: Carolyn A. Jost Robinson.

Ethnoprimatology Then, Ethnoprimatology Now

Like most "outside-of-the box" endeavors, ethnoprimatology's push to highlight the value of studying primates in anthropogenic contexts and develop research foci that extend across anthropology's subfields has not been without critique and resistance.[15] I recall when applying for grant funding to support my dissertation work on the ethnoprimatology of *Macaca tonkeana*, one of the reviewers commented that I was trying to do too much: I should choose to study the macaques or the people, but not both. The project would have lost its essence if I had chosen one or the other, so I did not give up: I did both (see Riley 2005). Similarly, in describing his experiences in publishing ethnoprimatological work, Agustín Fuentes (2011: 602) notes:

> Some reviewers met the chapters and essays with suspicion. Typical critiques included "These groups are in unnatural environments, you cannot truly test established ecological hypotheses" and "There is a lack of theory and hypothesis testing here. Where is the science?" This is the conundrum facing those attempting to publish ethnoprimatological work: one is forced to

navigate an opposition between nature and culture that emerges from the strict adaptationist/reductionist worldview typifying much of dominant primatology.

Nonetheless, I see tremendous value in this paradigmatic tension as it challenges us in the field of anthropological primatology to continue to re-evaluate and refine our concepts, constructs, and guiding frameworks. Since it first emerged in 1997, ethnoprimatology has undergone a process of maturation much like what zoologists Stuart Altmann and Jeanne Altmann (2003) describe for behavioral field studies in general. To date, a good portion of ethnoprimatological work has focused on proximate questions; this descriptive research has been critical for us to understand the types and characteristics of the various human-primate interfaces worldwide, and will remain so moving forward. But in recent years we have witnessed greater attention to "why" questions. As we saw in Chapter 4, recent theorizing in behavioral ecology and evolutionary biology (e.g., behavioral plasticity and niche construction) provides further theoretical guidance for ethnoprimatology. Today's and tomorrow's ethnoprimatology continues to value descriptive analysis, as any good field study should, but it is also theoretically rigorous. And, as we shall see in the next chapter, the theoretical guidance from which we can draw is not limited to the evolutionary sciences; recent theorizing on human-animal entanglements in the social sciences and humanities also has much to offer primatological practice (Fuentes 2012; Malone et al. 2014). In other words, we primatologists have more boundary dissolving to do, and by doing so, we will come to find, as our colleagues in sociocultural anthropology did long ago, that our theoretical landscape has room for more than one perspective (cf. Nader 2002). Ethnoprimatologists are also incorporating an increasingly diverse set of tools into their methodological repertoire (e.g., stable isotope analysis, new techniques in noninvasive sampling, social network analysis, participant observation, camera trapping), enabling us to ask new questions but also to ask old questions in new ways (see Dore et al. 2017; Loudon et al. 2014; Zak & Riley 2017).[16] Studying primates in the current "real world" necessitates theoretical and methodological pluralism (Riley 2018).

Notes

1 See Willis and colleagues (2004) perspective piece titled, "How 'virgin' is virgin rainforest?"

2 Orlove (1980: 235) broadly defines ecological anthropology as "the study of the relations among the population dynamics, social organization, and culture of human populations and the environments in which they live." The ecological anthropology of the 1960s and 1970s adopted an ecosystem approach from ecology and relied on key concepts such as adaptation, niche, carrying capacity to examine how human populations adapt to their environments though culture-based behavioral systems (Smith 1984; Moran

1990). Early ecological anthropology was criticized, however, for failure to recognize humans as social actors incurring culture change, and for its inability to *explain* cultural phenomena (Vayda & McCay 1975; Moran 1990; Brumfiel 1992). In the 1990s, a new kind of ecological anthropology emerged, "environmental anthropology," one that invoked a more multidimensional vision incorporating all aspects of the environment (physical, biological, social, and cultural), and one with greater attention to power relations and policy concerns (Brosius 1999; Kottak 1999).

3 "Primates in nature" is the title of primatologist Alison Richard's (1985) formative text on primate ecology and behavior, but also one that excludes any mention of humans as part of ecological environment.

4 See, for example, Fuentes and Wolfe (2002) and Paterson and Wallis (2005).

5 Even prior to Sponsel's 1997 paper, Shirley Strum and David Western (1982: 61) recognized the "need to consider primate ecology within a broader community framework which includes human activity." They went on to write: "Of equal significance is the tendency in primate studies...to either ignore or discount aspects of an animal's ecology influenced directly or indirectly by humans...[This] is specious in situations where are animal's ecology and behavior are significantly influenced by humans. That influence now extends to almost all natural settings" (p. 73).

6 Salience is a measure of how much knowledge respondents have and how important that knowledge is to them (Fleisher & Harrington 1998). I used salience scores as proxy measures of human forest use; that is, those resources identified as the most salient are assumed to be the most important and, thus, likely to be the most frequently sought after.

7 Matrix habitat generally refers to the anthropogenically-modified spaces that surround forest fragments (Gascon et al. 1999).

8 In some cases, however, evidence for a relationship between wild food availability and the timing of crop feeding/foraging remains equivocal. Geographer Naughton-Treves and colleagues (1998), for example, found that feeding on maize by three primate species at Kibale National Park, Uganda was not related to forest fruit availability.

9 It is also worth noting here the applied significance of the concept of fallback foods: improving conservation and management plans for nonhuman primates (Marshall et al. 2009). For example, efforts should be made to protect tree species that serve as fallback foods (e.g., figs) to potentially help buffer crop feeding, and conservation managers should consider planting fallback foods as parts of reforestation efforts (see Hockings et al. 2009; Marshall et al. 2009; McLennan 2013; Riley et al. 2013).

10 See Naughton-Treves et al. 1998; Hill 2000; Woodroffe et al. 2005; Strum 2010; Hill and Wallace 2012

11 See Daszak et al. (2001), Chapman et al. (2005), Wolfe et al. (2005), and Nunn and Altizer (2006). It is relevant here to critically point out that this multidisciplinary approach has almost exclusively incorporated researchers from the natural and health sciences (e.g., ecology, wildlife biology, conservation biology, veterinary medicine; see Daszak et al. 2001). More recently, though, researchers have recognized the importance of the anthropological approach for shedding light on the interspecies relationships that exist as a result of humans and wildlife living in ever-increasing proximity (e.g., Wolfe et al. 2005; McFarlane et al. 2012).

12 In late 2013, Jakarta's former governor, now president, Joko "Jokowi" Widodo banned performance monkeys (*topeng monyet*, i.e., masked monkeys) from Jakarta's streets in an effort to curb animal exploitation in the country (Anonymous 2013). This ban sparked other cities in Java to follow suit (e.g., Bandung; Rimadi 2013, Surakarta; Ayuningtyas 2013).

13 However, it would be incorrect to assume that all indigenous peoples who live in proximity to primates incorporate them into their cultural and symbolic domains. For example, Cormier (2002: 64) notes that while the Ka'apor and Guajá share ecological space and are linguistically related, in contrast to the Guajá, the Ka'apor do not prominently situate monkeys in their kinship and cosmology.
14 Tamil is a classical language from southern India.
15 Some of the critique is reflective of a more general resistance by natural scientists to heed any alternatives to or expand one's theoretical compass beyond the Neo-Darwinian paradigm (Ingold 2007).
16 See papers that were a part of the symposium "Expanded ecologies: Recent theoretical and methodological advances in the study of the human-primate interface" (www.asp.org/IPS/meetings/conferenceschedule.cfm?&parenteventid=6479) and the special issue titled "Ethnoprimatology in the 21st Century" in the *International Journal of Primatology* (October 2018, Volume 39, Issue 5).

References

Altmann SA, Altmann J. 2003. The transformation of behaviour field studies. *Animal Behaviour* 65: 413–23.

Anonymous. 2013. Jakarta governor Joko Widodo bans "abusive" monkey shows. In *www.news.com.au/travel/travel-updates/jakarta-governor-joko-widodo-bans-performing-monkeys/news-story/be9f7156eac339ca0612f3f37cefc3b3*: October 27, 2013

Ayuningtyas K. 2013. Surakarta to ban monkey shows from streets, November 7, 2013. In *The Jakarta Post*. www.thejakartapost.com/news/2013/11/07/surakarta-ban-monkey-shows-streets.html

Brosius JP. 1999. Analyses and interventions: Anthropological engagements with environmentalism. *Current Anthropology* 40: 277–310.

Brumfiel EM. 1992. Distinguished lecture in archeology: Breaking and entering the ecosystem—gender, class, and faction steal the show. *American Anthropologist* 94: 551–67.

Chapman CA, Gillespie TR, Goldberg TL. 2005. Primates and the ecology of their infectious diseases: How will anthropogenic change affect host-parasite interactions? *Evolutionary Anthropology* 14: 134–44.

Chapman CA, Wasserman MD, Gillespie TR, Speirs ML, Lawes MJ, et al. 2006. Do food availability, parasitism, and stress have synergistic effects on red colobus populations living in forest fragments? *American Journal of Physical Anthropology* 131: 525–34.

Constantino PJ, Wright BW. 2009. The importance of fallback foods in primate ecology and evolution. *American Journal of Physical Anthropology* 140: 599–602.

Cormier LA. 2002. Monkey as food, monkey as child: Guaja symbolic cannibalism. In *Primates Face to Face: Conservation Implications of Human-Nonhuman Primate Interconnections*, ed. A Fuentes, L Wolfe, pp. 63–84. Cambridge: Cambridge University Press.

Cormier LA, Urbani B. 2008. The ethnoprimatology of spider monkeys (*Ateles* spp.): From past to present. In *Spider Monkeys: Behavior, Ecology, and Evolution of the Genus Ateles*, ed. CJ Campbell, pp. 377–403. Cambridge: Cambridge University Press.

Cormier LA, Urbani B. 2015. The ethnoprimatology of the howler monkeys (*Alouatta* spp.): From past to present. In *Howler Monkeys: Behavior, Ecology, and Conservation*, ed. MM Kowalewski, PA Garber, Cortés-Ortiz, B Urbani, D Youlatos, pp. 259–80. New York: Springer.

Daszak P, Cunningham AA, Hyatt AD. 2001. Anthropogenic environmental change and the emergence of infectious diseases in wildlife. *Acta Tropica* 78: 103–16.

Dore KM, Riley EP, Fuentes A. 2017. *Ethnoprimatology: A Practical Guide to Research at the Human–Nonhuman Primate Interface.* Cambridge: Cambridge University Press.

Eley RM, Strum SC, Muchemi G, Reid GDF. 1989. Nutrition, body condition, activity patterns, and parasitism of free-ranging troops of olive baboons (*Papio anubis*) in Kenya. *American Journal of Primatology* 18: 209–19.

Engel G, Jones-Engel L, Schillaci M, Suaryana KG, Putra A, et al. 2002. Human exposure to herpesvirus B-seropositive macaques, Bali, Indonesia. *Emerging Infectious Diseases* 8: 789–95.

Fa JE. 1991. Provisioning of Barbary macaques on the Rock of Gibraltar. In *Primate Responses to Environmental Change*, ed. HO Box. London: Chapman & Hall.

Fleisher MS, Harrington JA. 1998. Freelisting: Management at a women's federal prison camp. In *Using Methods in the Field*, ed. VC deMunck, EJ Sobo, pp. 69–84. Walnut Creek, CA: Altamira Press.

Fuentes A. 2011. Measuring meaning and understanding in primatological and biological anthropology fieldwork: context and practice. In *Centralizing Fieldwork: Critical Perspectives from Biological and Social Anthropology*, ed. J MacClancy, A Fuentes, pp. 121–36. New York: Berghahn Books.

Fuentes A, Hockings KJ. 2010. The ethnoprimatological approach in primatology. *American Journal of Primatology* 72: 841–47.

Fuentes A. 2012. Ethnoprimatology and the anthropology of the human-primate interface. *Annual Review of Anthropology* 41: 101–17.

Fuentes A, Shaw E, Cortes J. 2007. Qualitative assessment of macaque tourist sites in Padangtegal, Bali, Indonesia, and the Upper Rock Nature Reserve, Gibraltar. *International Journal of Primatology* 28: 1143–58.

Fuentes A, Southern M, Suaryana KG. 2005. Monkey forests and human landscapes: is extensive sympatry sustainable for *Homo sapiens* and *Macaca fascicularis* on Bali? In *Commensalism and Conflict: The Human-Primate Interface*, ed. J Paterson, J Wallis, pp. 170–95. Norman, OK: American Society of Primatologists.

Fuentes A, Wolfe L, eds. 2002. *Primates Face to Face: The Conservation Implications of Human-Nonhuman Primate Interconnections.* Cambridge: Cambridge University Press.

Gascon C, Lovejoy TE, Bierregaard Jr RO, Malcolm JR, Stouffer PC, et al. 1999. Matrix habitat and species richness in tropical forest remnants. *Biological Conservation* 91: 223–29.

Graczyk TK, Bosco-Nizeyi J, Ssebide B, Thompson RCA, Read C, Cranfield MR. 2002. Anthropozoonotic *Giardia duodenalis* genotype (assemblage) A infections in habitats of free-ranging human-habituated gorillas, Uganda. *Journal of Parasitology* 88: 905–09.

Hardin R, Remis MJ. 2006. Biological and cultural anthropology of a changing tropical forest: A fruitful collaboration across subfields. *American Anthropologist* 108: 273–85.

Hill CM. 2000. Conflict of interest between people and baboons: Crop raiding in Uganda. *International Journal of Primatology* 21: 299–315.

Hill CM, Wallace GE. 2012. Crop protection and conflict mitigation: Reducing the costs of living alongside non-human primates. *Biodiversity and Conservation* 21: 2569–87.

Hockings KJ, Anderson J, Matsuzawa T. 2009. Use of wild and cultivated foods by chimpanzees at Bossou, Republic of Guinea: Feeding dynamics in a human-influenced environment. *American Journal of Primatology* 71: 636–46.

Ingold T. 2007. The trouble with 'evolutionary biology'. *Anthropology Today* 23: 13–17.

Jaman MF, Huffman MA. 2013. The effect of urban and rural habitats and resource type on activity budgets of commensal rhesus macaques (*Macaca mulatta*) in Bangladesh. *Primates* 54: 49–59.

Jones-Engel L, Engel GA, Schillaci MA, Babo R, Froehlich J. 2001. Detection of antibodies to selected human pathogens among wild and pet macaques (*Macaca tonkeana*) in Sulawesi, Indonesia. *American Journal of Primatology* 54: 171–78.

Jones-Engel L, Schillaci M, Engel G, Paputungan U, Froehlich J. 2005. Characterizing primate pet ownership in Sulawesi: Implications for disease transmission. In *Commensalism and Conflict: The Human-Primate Interface*, ed. JD Paterson, J Wallis, pp. 196–221. Norman, OK: American Society of Primatologists.

Jones-Engel L, Engel GA, Schillaci MA, Froehlich J, Paputungan U, Kyes RC. 2004. Prevalence of enteric parasites in pet macaques in Sulawesi, Indonesia. *American Journal of Primatology* 62: 71–82.

Kottak CP. 1999. The new ecological anthropology. *American Anthropologist* 101: 23–35.

Lambert JE, Chapman CA, Wrangham RW, Conklin-Brittain NL. 2004. Hardness of cercopithecine foods: Implications for the critical function of enamel thickness in exploiting fallback foods. *American Journal of Physical Anthropology* 125: 363–68.

Lane KE, Holley C, Hollocher H, Fuentes A. 2011. The anthropogenic environment lessens the intensity and prevalence of gastrointestinal parasites in Balinese long-tailed macaques (*Macaca fascicularis*). *Primates* 52: 117–28.

Lane KE, Lute M, Rompis A, Wandia IN, Arta Putra IGA, et al. 2010. Pests, pestilence, and people: The long-tailed macaque and its role in the cultural complexities of Bali. In *Indonesian Primates*, ed. S Gursky-Doyen, J Supriatna, pp. 235–48. New York: Springer.

Lizzaralde M. 2002. Ethnoecology of monkeys among the Bari of Venezuela: perception, use and conservation. In *Primates Face to Face: Conservation Implications of Human-Nonhuman Primate Interconnections*, ed. A Fuentes, L Wolfe, pp. 85–100. Cambridge: Cambridge University Press.

Loudon JE, Grobler JP, Sponheimer M, Moyer K, Lorenz JG, Turner TR. 2014. Using the stable carbon and nitrogen isotope compositions of vervet monkeys (*Chlorocebus pygerythrus*) to examine questions in ethnoprimatology. *PloS one* 9: e100758.

Loudon JE, Sauther ML, Fish DD, Hunter-Ishikawa M, Ibrahim YJ. 2006. One reserve, three primates: Applying a holistic approach to understand the interconnections among ring-tailed lemurs (*Lemur catta*), Verraux's sifaka (*Propithecus verrauxi*), and humans (*Homo sapiens*) at Beza Mahafaly Special Reserve, Madagascar. *Ecological and Environmental Anthropology* 2: 54–74.

Malone N, Wade AH, Fuentes A, Riley EP, Remis MJ, Jost Robinson CA. 2014. Ethnoprimatology: Critical interdisciplinarity and multispecies approaches in anthropology. *Critique of Anthropology* 34: 8–29.

Maples WR, Maples MK, Greenhood WF, Walek ML. 1976. Adaptations of crop-raiding baboons in Kenya. *American Journal of Physical Anthropology* 45: 309–16.

Marshall AJ, Boyko CM, Feilen KL, Boyko RH, Leighton M. 2009. Defining fallback foods and assessing their importance in primate ecology and evolution. *American Journal of Physical Anthropology* 140: 603–14.

Marshall AJ, Wrangham R. 2007. Evolutionary consequences of fallback foods. *International Journal of Primatology* 28: 1219–35.

McFarlane R, Sleigh A, McMichael T. 2012. Synanthropy of wild mammals as a determinant of emerging infectious diseases in the Asian–Australasian region. *EcoHealth* 9: 24–35.

McGrew WC. 2007. New wine in new bottles: Prospects and pitfalls of cultural primatology. *Journal of Anthropological Research* 63: 167–82.

McKinney T. 2011. The effects of provisioning and crop-raiding on the diet and foraging activities of human-commensal white-faced capuchins (*Cebus capucinus*). *American Journal of Primatology* 73: 439–48.

McLennan MR. 2013. Diet and feeding ecology of chimpanzees (*Pan troglodytes*) in Bulindi, Uganda: Foraging strategies at the forest-farm interface. *International Journal of Primatology* 34: 585–614.

Moran EF. 1990. *The Ecosystem Approach in Anthropology: From Concept to Practice.* University of Michigan Press.

Nader L. 2002. Missing links: A commentary on Ward H. Goodenough's moving article "Anthropology in the 20th century and beyond". *American Anthropologist* 104: 441–49.

Naughton-Treves L, Treves A, Chapman C, Wrangham R. 1998. Temporal patterns of crop-raiding by primates: Linking food availability in croplands and adjacent forest. *Journal of Applied Ecology* 35: 596–606.

Nekaris KAI, Shepherd CR, Starr CR, Nijman V. 2010. Exploring cultural drivers for wildlife trade via an ethnoprimatological approach: A case study of slender and slow lorises (*Loris* and *Nycticebus*) in South and Southeast Asia. *American Journal of Primatology* 72: 877–86.

Nunn C, Altizer SM. 2006. *Infectious Diseases in Primates: Behavior, Ecology and Evolution.* Oxford: Oxford University Press.

Orlove BS. 1980. Ecological anthropology. *Annual Review of Anthropology* 9: 235–73.

Paterson JD, Wallis J, eds. 2005. *Commensalism and Conflict: The Human-Primate Interface*, Vols. 4. Norman, OK: American Society of Primatologists.

Radhakrishna S. 2018. Primate tales: Using literature to understand changes in human-primate relations. *International Journal of Primatology* 105: 230–37.

Reinhard J. 1988. The Nazca lines, water and mountains: an ethno-archaeological study. In *Recent Studies in Pre-Columbian Archaeology*, ed. N Saunders, O de Montmollin, pp. 363–414. Oxford: British Archaeological Reports.

Richard AF. 1985. *Primates in Nature.* WH Freeman and Co. New York.

Riley EP. 2005. *Ethnoprimatology of Macaca tonkeana: The Interface of Primate Ecology, Human Ecology, and Conservation in Lore Lindu National Park, Sulawesi, Indonesia.* Ph.D. Dissertation thesis. University of Georgia.

Riley EP. 2007. The human-macaque interface: Conservation implications of current and future overlap and conflict in Lore Lindu National park, Sulawesi, Indonesia. *American Anthropologist* 109: 473–84.

Riley EP. 2018. The maturation of ethnoprimatology: Theoretical and methodological pluralism. *International Journal of Primatology* 39: 705–29.

Riley EP, Priston NEC. 2010. Macaques in farms and folklore: Exploring the human-macaque interface in Sulawesi, Indonesia. *American Journal of Primatology* 72: 848–54.

Riley EP, Tolbert B, Farida WR. 2013. Nutritional context explains the attractiveness of cacao to crop raiding Tonkean macaques. *Current Zoology* 59: 160–69.

Rimadi L. 2013. Walikota Bandung Ikuti Jokowi Tertibkan Topeng Monyet. *http://news.liputan6.com/read/732822/walikota-bandung-ikuti-jokowi-tertibkan-topeng-monyet*

Saj T, Sicotte P, Paterson JD. 1999. Influence of human food consumption on the time budget of vervets. *International Journal of Primatology* 20: 977–94.

Schillaci M, Jones-Engel L, Engel G, Paramastri Y, Iskandar E, et al. 2005. Prevalence of enzootic simian viruses among urban performance monkeys in Indonesia. *Tropical Medicine and International Health* 10: 1305–14.

Sha JCM, Hanya G. 2013. Diet, activity, habitat use, and ranging of two neighboring groups of food-enhanced long-tailed macaques (*Macaca fascicularis*). *American Journal of Primatology* 75: 581–92.

Shepard GH. 2002. Primates in Matsingenka subsistence and world view. In *Primates Face to Face: Conservation Implications of Human-Nonhuman Interconnections*, ed. A Fuentes, L Wolfe, pp. 101–36. Cambridge: Cambridge University Press.

Siex KS, Struhsaker TT. 1999. Ecology of the Zanzibar Red Colobus monkey: demographic variability and habitat stability. *International Journal of Primatology* 20: 163–91.

Smith EA. 1984. Anthropology, evolutionary ecology, and the explanatory limitations of the ecosystem concept. In *The Ecosystem Concept in Anthropology*, pp. 51–85. Boulder, CO: Westview Press.

Sponsel LE. 1997. The human niche in Amazonia: Explorations in ethnoprimatology. In *New World Primates: Ecology, Evolution, and Behavior*, ed. WG Kinnzey, pp. 143–65. New York: Aldine Gruyter.

Struhsaker TT. 1997. *Ecology of an African Rain Forest: Logging in Kibale and the Conflict between Conservation and Exploitation*. Gainesville, FA: University of Florida Press.

Strum SC. 1994. Prospects for management of primates pests. *Revue D'Ecologie (La Terre et La Vie)* 49: 295–306.

Strum SC. 2010. The development of primate raiding: Implications for management and conservation. *International Journal of Primatology* 31: 133–56.

Strum SC, Western JD. 1982. Variations in fecundity with age and environment in olive baboons (*Papio anubis*). *American Journal of Primatology* 3: 61–76.

Vayda AP, McCay BJ. 1975. New directions in ecology and ecological anthropology. *Annual Review of Anthropology* 4: 293–306.

Vellend M. 2010. Conceptual synthesis in community ecology. *The Quarterley Review of Biology* 85: 183–206.

Ward C. 2003. The evolution of human origins. *American Anthropologist* 105: 77–88.

Willis KJ, Gillson L, Brncic TM (2004) How "virgin" is virgin rainforest? Science 304: 402–403.

Wheatley B. 1999. *The Sacred Monkeys of Bali*. Prospect Heights, IL: Waveland Press, Inc.

Wolfe ND, Daszak P, Kilpatrick AM, Burke DS. 2005. Bushmeat hunting, deforestation, and prediction of zoonotic disease. *Emerging Infectious Diseases* 11: 1822.

Woodroffe R, Thirgood S, Rabinowitz A, eds. 2005. *People and Wildlife: Conflict or Coexistence?* Cambridge: Cambridge University Press.

Wright PC, Andriamihaja BR, Raharimiandra SA. 2005. Tanala synecological relations with lemurs in southeastern Madagascar. In *Commensalism and Conflict: The Human-Primate Interface*, ed. J Paterson, J Wallis, pp. 119–45. Norman, OK: ASP.

Zak AA, Riley EP. 2017. Comparing the use of camera traps and farmer reports to study crop feeding behavior of moor macaques (*Macaca maura*). *International Journal of Primatology* 38: 224–42.

6

BEYOND THE DIVIDE

Fieldwork, Reflexivity, and Multispecies Worlds

As we saw in Chapter 2, over a half a century ago Boas was concerned about the rift between the biological and cultural threads of anthropology. Internal debates concerning what anthropology is or should be have continued over the years, and recently there have been attacks on anthropology's broader merit outside the academic realm.[1] It is therefore a better time than ever for anthropological primatologists to reflect upon our field. An important part of doing so is accepting the reality that it is becoming increasingly difficult to encounter a primate population free from human influence. Accordingly, as I have argued in Chapters 4 and 5, moving forward, the anthropogenic element is likely to be relevant to every primatological study. Our domain of study is unquestionably multispecies, and one of those species is the complex and complicated: *Homo sapiens*. We can choose to ignore this dimension – to the detriment of our field, I would argue – or we can look beyond the borders of traditional primatological practice. As anthropologist Marcus Baynes-Rock (2013:223) so elegantly relays in his study of the shared ecology of hyenas and humans in Ethiopia, "multispecies commons," that is, zones of entanglement where species come together, produce such complexity that ethological observations *alone* are not enough. In the historic city of Harar, hyenas and humans are involved in mutual co-shaping (cf. Fuentes 2010a) – the hyenas come to feeding sites within the walls of the town which benefits tourism and the practice and location of the feeding shapes the territorial boundaries between hyena clans – such that "the biological is the social is the historical is the ecological" (Baynes-Rock 2013: 224). Multispecies commons comprising human and nonhuman primates are undoubtedly equally complex. Primatology itself emerged from the crossing of disciplinary boundaries (Rowell 2000); today, existing boundaries are being redefined and new boundaries are being crossed and blurred.

In this chapter, I intend to show the insight that contemporary primatology gains from re-engaging with other subfields of anthropology (e.g., sociocultural anthropology, environmental anthropology) and other social sciences (e.g., sociology, human geography) to study the multispecies contexts that characterize the lives of most contemporary primate populations. In doing so, I explore: How can reaching across disciplinary divides improve traditional primatological practice? And, in doing so, how is the scope of anthropological primatology expanded beyond its original purpose of elucidating human biological evolution?

Sociocultural Anthropology: Friend Not Foe

Like primatology, sociocultural anthropology has a multidisciplinary parentage; but, it has a much deeper and wider history that has witnessed the emergence of a multitude of paradigms aimed at deciphering why humans do what they do, what culture is, and what it means to people. Accordingly, sociocultural anthropology is concerned with all facets of humankind, focusing on our similarities and differences cross-culturally. Sociocultural anthropologists study, for example, the political and economic structures that humans create and that empower and constrain people across different scales; the dynamic ways that humans construct identities; the ways in which humans interface with their ecological environments, and how culture shapes people's approaches to health, disease, and illness. And, while certain paradigms have held reign over others across the years, sociocultural anthropology has never been guided by just one theoretical framework, as traditional primatology has with evolutionary theory.

So, what does sociocultural anthropology have to do with primatology and how can engaging with it make primatology better? To address these questions, a good place to start is by looking at important points of convergence between sociocultural anthropology and primatology. For example, both fields have challenged dualistic thinking. Anthropological primatology employs a broad comparative approach that emphasizes commonality and continuity across the primate Order (Strier 2003), thereby challenging human/animal dualism. Sociocultural anthropology also sees human-animal dualism as problematic, arguing instead for human-animal interconnectedness (Knight 2005a) and that being "human" means being with other animals.[2] Second, as we learned in Chapter 3, the methodological roots of primatology are situated in sociocultural anthropology: early anthropological primatologists were trained in the tradition of producing rich, descriptive accounts of behavior. These accounts derived from diary-style recorded observations and detailed anecdotes from the field. Due to the development of ethological methods in biology in the late 1960s and the 1970s (e.g., Altmann 1974), however, primatology's methodology shifted toward an emphasis on systematic behavioral sampling and quantitative analysis. This quantitative approach continues to reign. Nonetheless, primatology continues to share a key methodological affinity with sociocultural anthropology: the tradition of fieldwork.[3]

By conducting situated and engaged fieldwork, primatologists and sociocultural anthropologists alike have always been and will always be participant observers (Strier 2003; Knight 2017). During the process of habituation, for example, primatologists (like ethnographers) "work hard to establish rapport and win the trust of their subjects," carefully monitoring how our study subjects "come to regard us" (Arluke & Sanders 1996: 22).[4] After a certain level of habituation is achieved, primatologists typically view themselves as now primed to collect "real data." However, like ethnographers, primatologists never really become a truly neutral element or are completely accepted by our primate study subjects; rather, we (primatologists) become "member enough" (Arluke & Sanders 1996: 23; see also Hanson & Riley 2018). Primatologists may strive to remain detached during the habituation process, but in reality, like ethnographers, what we are doing is seeking trust from our informants, albeit nonhuman. Moreover, habituation is not merely an outcome, it is an intersubjective process; a process during which humans and primates read and come to know one another (Hanson & Riley 2018; Figure 6.1).

Primatology also shares an epistemological affinity with sociocultural anthropology: reflexivity. There, I said it. Reflexivity – understanding and managing our

FIGURE 6.1 Staff members at the Uluwatu temple enjoy a coffee break while habituated long-tail macaques (*Macaca fascicularis*) and a dog look on, Bali, Indonesia. Photo credit: Jeffrey V. Peterson.

influence on the research process – tends to (unfortunately) trigger considerable disdain among some primatologists.[5] And yet, the field of primatology has benefitted tremendously from this humanist tradition. Do we need to call it something else so we can appreciate that? Maybe. How about *subjective objectivity*? As part of the debate about "anthropology as science" that unfolded in *Anthropology News* in the mid-1990s, anthropologist Liza Cerroni-Long (1996) challenges what she sees as an artificial dichotomy between science and humanities, arguing that "the practice of science is a messy, bumbling, and all-to-human affair" and that the production of anthropological knowledge can only be built on "subjective objectivity" (p. 52). In other words, anthropologists calibrate their findings via reflexivity and enrich and round out the content of their research with rich description of the context. Regardless of what we call it, we (primatologists) are well served by opening up the blinders so that we can see how reflexivity/subjective objectivity benefits primatological practice.

The Ethological Gaze: Seeing Reflexivity in Primatology

Emerging as part of a "reinventing" of anthropology (Hymes 1972), reflexivity is the constant awareness and consideration of how we as researchers shape the research process, the data we collect, and the conclusions we make.[6] Being reflexive means that we are aware of the contingencies of knowledge (Boyer 2015) and that "gender, race, class, sexual, and other positions cultivate affinities for some questions over others" (Lichterman 2015: 2). In short, positionality matters. When doing ethnography, socicocultural anthropologists talk to and share space with people; but they also observe, or see, people. In so doing, they must be cognizant of what they see and what they don't see. As anthropologist Raymond Madden (2010: 101) notes, "being reflexive about one's 'way of seeing' is crucial to the development of [one's] 'ethnographic gaze'," or how we train our eye to look and see. Just as one's ethnographic gaze is reflective of our positionality, I would argue that so too is our "ethological gaze." In other words, if anthropological inquiry is "culturally mediated" and "never only scientific" (Scholte 1972: 431), then so too is anthropological primatological inquiry. Indeed, cultural studies of science are more typically done by those from the outside looking in, with the aim of not "telling working scientists anything they don't already know" but rather that their "description of the way in which [the scientists'] craft activities become transformed into 'statements about science' might constitute a new perspective on what working scientists know to be the case" (Latour & Woolgar 1979: 30).[7] But, it can also be done from within (primatology), and it has.

It is important to note that, unlike in ethnography for which reflexivity is integrated into the research process and writing, in primatology, being reflexive has largely been done post-hoc – that is, scholars have reflected on how their biases have influenced the kinds of questions they ask (and the answers provided), the behaviors they chose to observe, and their overall approach to research *after the fact.*[8] This is not meant as a critique though, for this reflexivity has, in turn, helped shape the direction and focus of subsequent research. For example, the predominant ethological gaze during anthropological primatology's early years saw primate males as the stabilizing force in primate society (e.g., DeVore 1965). Early anthropological primatologists believed in the "naturalness" of the male dominance hierarchy and it was assumed that dominance *always* equals reproductive success (Strum 1982). In the late 1970s and early 1980s, however, scholars began to criticize DeVore's baboon model, noting that it reflected Western ideas of how the world should be: a division of male and female roles, male political power, and female domesticity (Strum 1987a).[9] Reflexive awareness of Western influences and continued fieldwork on baboon society by other researchers identified the importance of social strategies, such as friendships, in baboon society, with the key to success being "social finesse, not brutal force" (Strum 1987b: 689; Figure 6.2). These critiques led to a shifting lens regarding the centrality of aggression in primate society toward consideration of other (equally) important components of primate social lives such as cooperation (Sussman et al. 2005), reconciliation (de Waal 1989), and the importance of social tolerance and affiliation not just among females but also among males (Hill & van Hooff 1994).

Primatology in the 1970s and 1980s witnessed a growing research focus on females, largely due to continued self-reflection influenced by new feminist perspectives and an influx of women into the field.[10] These studies challenged

FIGURE 6.2 Baboon male with female friend and her infant. Notice how relaxed the mother is despite the male holding the infant in the "wrong" way. Uaso Ngiro Baboon Project, Laikipia, Kenya. Photo credit: Shirley C. Strum.

the male-centrism of earlier work, demonstrating that females are not merely passive objects of mate competition and that females form the core structure in many primate societies. Still today, we continue to reveal the significance of the diverse roles female primates play. For example, long-term research on northern muriquis (*Brachyteles hypoxanthus*) in Brazil has identified the importance of lifelong maternal investment for the reproductive success of co-resident sons (Strier et al. 2011) and recent fieldwork on bonobos (*Pan paniscus*) – another male philopatric species – has further elucidated the patterning of female-female coalitions, with older females supporting younger females to prevent harassment from males (Tokuyama & Furuichi 2016).

Reflexivity has also led primatologists to consider how their own sociocultural contexts shape ideas of what is "typical" (Strier 1994). In modern Judeo-Christian worldview, monogamy constitutes what is typical for human primates. Accordingly, the hylobatids (gibbons and siamangs), long heralded as the quintessential pair bond, the model monogamists if you will, were considered particularly worthy of in-depth study for the insight they could provide on the evolution of monogamy in the human species (Fuentes 2010b). Field study after field study, however, began to reveal that hylobatids can, and do, live in groups comprising more than two adults and engage in extragroup copulations, and hence, exist in "variable communities" rather than in typical pair-bonds (Fuentes 2000: 51).[11] Reflection on these findings, in turn, led scholars to rethink what "monogamy" really is – that is, a mating pattern rather than a social system (Fuentes 2000).

Similarly, in anthropologist Pamela Asquith's (1991, 1996, 2000) in-depth study of Western and Japanese traditions of primatology, we see that the lack of a Christian worldview (which sees humans as separate from, and privileged over, nature and the only beings to possess souls) during the time primatology was developing in Japan meant that Japanese primatologists had no qualms about attributing motivation, feeling, and personality to individual primates.[12] Therefore, differing socio-religious worldviews explain, in part, the differences in the methodological approaches of the two traditions. The Japanese tradition emphasized the importance of individual identification, embraced the use of anthropomorphism and descriptive accounts of behavior, and relied heavily on provisioning to enable close observations of individuals (Takasaki 2000; de Waal 2003). By seeking to know individuals, attributing motive to their actions, and eschewing the strictly biological explanations of behavior offered by the sociobiological perspective (embraced in West in the 1970s and onward), Japanese primatologists were able to "see more" (Asquith 1996: 253) – and importantly, as Asquith (1996) notes, they were able to see aspects of behavior that only later Western primatologists came to appreciate. For example: the critical observation that male dominance is not merely a function of strength, but also a function of the friendships that males build with females (Itani 1961), and the importance of kinship, a supposedly strictly human phenomenon, in primate societies (Reynolds 1992).

Recognizing and attending to our biases is part of the reflexive critique that enables us to appreciate how culture is at work in the practice of science (Nader 2001: 610). The rise of feminism and ensuing "reconstructions of sex and gender in primate story telling" (Haraway 1989: 301; see also Strum & Fedigan 2000) and the power of socio-religious tradition in influencing our observations and interpretations demonstrate that we are shaped by the broader social and political contexts in which we work. The abovementioned examples also demonstrate the invaluable role that accumulated fieldwork plays in producing and molding knowledge and understanding (Fuentes 2010b). In her sociological study of the infanticide controversy in primatology, philosopher of science Amanda Rees (2009) discusses what she calls the "fieldworker's regress" – that is, how "field researchers are constantly called on to reflexively evaluate their own observations and the conditions under which they were made" (pp. 9–10); field primatologists are no exception to this.[13] When we recognize that reflexivity combined with a fundamental tradition of fieldwork characterizes *both* primatology and socicocultural anthropology, Boas' concerns of a splintered field are assuaged and more recent claims of a wide epistemological abyss between the two fields become unfounded.[14]

Context Always Matters, So Study It

Another reason primatologists benefit from reaching across the divide to sociocultural anthropology is that we are afforded the opportunity to understand the social, cultural, economic, and political contexts that affect the lives and livelihoods of primate populations. When we understand these contexts, we can better interpret the observed patterns of primate behavior. Certainly, as we learned in Chapter 5, understanding the anthropogenic context is an integral component of the ethnoprimatological approach, but my point here is that it is critical to *all* primate field studies, however apparent or unapparent the anthropogenic footprint is.[15] To illustrate how context matters I provide examples from my own research in Sulawesi, Indonesia and the Silver River, Florida, and anthropologists Agustín Fuentes' and Kerry Dore's work in Bali, Indonesia and St. Kitts, respectively.

The Human-Macaque Community Interface in the Lindu Highlands

For my dissertation research, I chose a field site, for better or worse, where no primate research had ever been done – the Lindu highlands of Central Sulawesi, Indonesia. After visiting a number of locations around Lore Lindu National Park – one of the largest National Parks in Sulawesi, comprising a total area of 217,982 ha – I settled on the idyllic Lake Lindu highland plain as my research site. The Lake Lindu area, situated at approximately 1000 m.a.s.l. is one of two valley enclaves that are allowed to exist within the National Park because it is a major rice growing area and has long established settlements. Indeed, I chose this site because I was explicitly interested in understanding the ecological and cultural interconnections between villagers and Tonkean macaques; however, I could have easily arrived at this site solely to investigate the behavioral ecology of Tonkean macaques without considering the human dimension. Had I done so, I would be without context to fully understand my data. Namely, it was critical that I understood the primates *and* the people. For the humans, I needed to understand what their primary livelihood strategies were, and how they perceived the National Park, and its resources and functions.

The people indigenous to the Lake Lindu plain are members of the Kaili ethnic group, which is further divided into seven distinct groups on the basis of dialect, with the Lindu form referred to as Kaili Tado' (Acciaioli 1989). Despite linguistic similarity, the Lindu highlanders were converted to Christianity, and hence, do not call themselves Kaili; instead, they reserve this term for their lowland, Muslim neighbors in the Palu valley and Parigi, and call themselves *To Lindu* (people of Lindu) (Davis 1976). Dutch conquest of the Kaili highlands began in 1902, and at Lake Lindu resulted in the concentration of hill farmers into three settlements (i.e., Anca, Tomado, and Langko) along the lakeside (Acciaioli 1989; Li 2000). Throughout the colonial and post-independence eras the Lindu plain remained a relatively isolated area. Since that time, however, it has experienced considerable in-migration, including other Kaili from nearby lowland towns who migrated to the area under local resettlement schemes (Acciaioli 1989) and spontaneously migrating Bugis people from South Sulawesi, who were, and continue to be, attracted to the area for wet-rice agriculture, perceived available land for the planting of important cash crops, and the development of a fishing industry at the 3,000 ha lake.[16]

Studying the human community necessitated ethnographic methods: I learned Bahasa Indonesia (the national language of Indonesia) *in-situ*; I lived within the community throughout the duration of the study, fully immersing myself into "life in Lindu"; and I employed classic ethnographic techniques such as participant observation (e.g., with cacao farmers to learn about their planting and harvesting practices and how they mitigate crop feeding by wildlife) and in-depth interviews and focus groups with village members. Through this approach, I learned that the local community of Tomado, the village in which I lived, was by no means homogenous, and that the existing diversity in backgrounds (i.e., ethnicities, lifestyles) could translate into differing views, concerns, and interests with regard to forest use and protection (see Ostrom 1999).[17] And, indeed, I found that there was considerable diversity in knowledge, attitudes, and perceptions of the forest and its conservation that varied by cultural-ecological and ethnic identity. While the Lindu recognized the importance of the protected area to guard resources for the future, migrants often saw it solely in negative terms, as limiting their ability to expand their agricultural enterprises (Riley 2013). I also learned that *To Lindu* folk ecology includes stories about human-macaque interactions. Namely, many Lindu envision themselves linked to macaques via folklore that describes a particular tale of a human-macaque interaction that simultaneously imbues respect and contempt for the macaques (Riley 2010). For some Lindu this cultural interconnection translates into tolerance as they encounter the macaques in their shared ecological space (e.g., if macaques feed on their crops). Knowing this helped explain why I was able to successfully habituate some macaque groups but not others despite their living in similar matrix habitat; it mattered who managed the land. For example, the ranges of two groups (that I spent months trying to habituate with minimal success) included agricultural land managed by migrants who did not tolerate macaque crop feeding behavior; rather, they retaliated against it by chasing the macaques and throwing rocks at them.

I also found that an understanding of people's livelihood strategies was critical to evaluating patterns of macaque crop feeding, habitat use, and the long-term survivability of certain groups. During the colonial era, the indigenous *To Lindu* subsisted primarily on *ladang* (swidden) agriculture, consisting of maize, tubers, and other dry-crops. At the time of my research (2002–2004), wet-rice agriculture (*sawah*) predominated in Lindu, and was practiced by both indigenous Lindu and migrants, but tree cash crops, such as coffee and cacao, which were introduced by Bugis migrants in the 1970s, had become an important part of the Lindu economy. However, in early 2003, one year after I arrived to conduct my research, a major shift in many individual's primary source of subsistence from farming to fishing occurred throughout the Lindu enclave due to the population explosion of profitable tilapia fish in the lake (Riley 2007a). This economic livelihood shift in turn resulted in ecological change across people's shared space with macaques, including the felling of many large old-growth hardwood species for dug-out canoes. It also led some farmers to completely abandon their cacao crops. These outcomes likely had important impacts on the behavioral ecology and implications for survivability of neighboring groups of Tonkean macaques. For example, I noted that within the span of three months in early 2003, 15 canopy-sized trees were felled in the core home range area of one of my study groups, for the sole purpose of constructing dug-out canoes for use in the burgeoning fishing industry (Figure 6.3). Three of those trees were known to be specific food trees for this macaque group, while the remainder (particularly, *Palaquium obovatum*) provided other resources for the macaques, such as canopy connectivity and strata for resting and foraging. A burgeoning fishing industry also meant that some farmers chose to focus their energy on fishing to the neglect of their cacao gardens. This socioeconomic shift meant that gardens were no longer being maintained, and importantly, no longer being guarded. Had I been following any of the macaque groups whose ranges included these gardens, I would have likely observed shifts in their ranging and foraging patterns, which only would have made sense in context.

I also learned that knowing villagers' resource use patterns and folklore was critical to understanding the density and distribution patterns of macaque food resources. Documenting the structure and composition of primate habitats,

FIGURE 6.3 Villagers carve out a dug-out canoe from a felled tree, Anca, Lindu, Central Sulawesi, Indonesia. Photo credit: Rusmanto Lakareba.

including the species richness, density, and distribution of key food resources, is an integral component of primate behavioral ecology studies. But in spaces where humans are in the mix, we also need to know how human behavior may shape these ecological features. Take, for example, a species of palm ubiquitous to Southeast Asia: the sugar palm (*Arenga pinnata*). People actively manage its cultivation largely because it is regarded as one of the most important forest resources due to the diverse array of products it generates (Davis 1988; Mogea et al. 1991). As a result, the palm species thrives in human-altered habitats. As I explained in Chapter 5, this palm also provides an important food source for Tonkean macaques. So, in this case, we see human resource use and management shaping the structure of Tonkean macaque habitat, and in turn, influencing the foraging patterns that we (as primatologists) observe. A second example involves a particularly important resource for many species of tropical frugivorous vertebrates – fig fruit (*Ficus* spp.) – and folklore surrounding it. Recognized as "keystone species" in some areas of the world, figs are important for wildlife, particularly birds and primates, due to the fact that their asynchronous fruiting patterns make them a reliable food source when other resources are scarce (Kinnaird & O'Brien 2005).[18] In Lindu, I found that fig fruit is important to the diet of Tonkean macaques living in both heavily-altered and less-altered forests, comprising 22 percent and 51 percent of their plant species diet, respectively (Riley 2007b). Although human forest product use and conversion of forest to agricultural spaces in Lindu has resulted in lower overall key food density for Tonkean macaques (Riley 2007b), the felling of fig trees is historically an uncommon occurrence in Lindu, even in permitted use areas, primarily due to local folklore contending that strangling fig trees house evil spirits that should not be disturbed. These examples demonstrate the importance of examining how humans use resources in order to understand why those resources persist in human-modified landscapes and the role they may play in ensuring the long-term persistence of primates.

Natural-Cultural Contact Zones in Bali, Indonesia

The Indonesian island of Bali presents another setting in which humans and other primates have a long history of mutual co-shaping of their shared ecological and social spaces (Wheatley 1999; Fuentes 2010a). While their sympatry dates back for millennia, the last several hundred years have witnessed tremendous human population growth and increasingly intensified agricultural systems characterized by irrigated landscapes regulated through a complex web of water temples (Lansing 1991). These feats of human ecological engineering have resulted in a patchwork of shrines and temples amidst mosaics of riverine forests that, in turn, have shaped the ranging and dispersal patterns, population genetics, and disease ecology of co-autochthonous long-tailed macaques (*Macaca fascicularis*) (Fuentes et al. 2005; Lane et al. 2010, 2011). Moreover, the ritual behavior that takes place at the temple sites, such as the daily placement of offerings, which the macaques feed on, affects macaque foraging and feeding strategies (Fuentes 2010a; Figure 6.4). While mutual tolerance typically prevails at temple sites during ritual behavior, the agricultural land that lies beyond becomes arenas where resources are contested and conflict often ensues as macaques attempt to consume cultivated foods and farmers retaliate. Many of these "naturalcultural contact zones" (Fuentes 2010a: 607; see also Haraway 2003) have also become tourist sites, thereby adding to the complexity of the human dimension and social niches constructed (Figure 6.5). Accordingly, people's perceptions of the Bali macaques are diverse: they are, paraphrasing Fuentes (2010a: 611), crop feeding pests, tolerated co-residents, participants in Balinese Hindu rituals, and tourist attractions.

Primates in Foreign Lands

Context also proves to be important when investigating primate populations that live in areas outside of their native distribution. While there is evidence that primates themselves are able to extend their range beyond their original distribution,[19] a number of primate species currently exist outside of their native zone due to the process of ethnophoresy, the dispersal of animals to other regions through

FIGURE 6.4 Long-tailed macaques (*Macaca fascicularis*) at the Monkey Forest in Ubud, Bali, Indonesia, feed on offerings left for them during the Tumpek Kandan ceremony. Photo credit: Atrifary Juita.

introduction or translocation (often deliberate) by humans (Heinsohn 2003; see also Gumert et al. 2011; Dore 2017). A case in point: the rhesus macaques (*Macaca mulatta*) of Florida's Silver River. Primates have, of course, been extinct from the United States for millions of years, but a population of rhesus macaques has been living in the riparian woodlands of north central Florida for over 80 years (Figure 6.6). One cannot even begin to understand the demography and behavioral ecology of this population without considering the human dimension, for it is because of humans that the population lives there in the first place. Although there are no known written reports on the macaques' initial introduction, local folklore maintains that a number of individuals were released by a tour boat operator by the name of Colonel Tooey who was eager to enhance the excitement value of his Jungle Cruise tours. In order to keep the macaques close to the river's edge for easy viewing by tourists, tour boat operators, and later the management of the attraction at the river's headsprings, provisioned them for decades (Maples et al. 1976).[20] By the 1980s, however, the Silver River macaques' existence along the river had become increasingly controversial. The Florida Game and Fresh Water Fish Commission (FGFWFC) was concerned about the growing population size and potential impact of this exotic species on native ecological communities, in addition to the possibility of disease transmission. As a result, the FGFWFC initiated efforts to control the population, including a sterilization program and trapping and removal.[21] The citizens of Marion Country, Florida, however,

FIGURE 6.5 Visitors observe long-tailed macaques (*Macaca fascicularis*) eating provisioned food at Uluwatu temple, Bali, Indonesia. Photo credit: Jeffrey V. Peterson.

strongly favored their continued existence along the river, some of whom rallied together to form a group known as the "Friends of the Silver River Monkeys" which secured 25,000 signatures from local residents in protest of the monkeys' removal from public lands. This historical tracing demonstrates that human actions and the meaning humans give to the macaques have, from the very start, shaped the macaques' population demographics, and in turn, the structure and organization of its social groups, group ranging patterns and foraging strategies, and overall ability to colonize a new world.

In 2013 I was fortunate to join a collaborative research team, comprising primatologists, sociocultural anthropologists, and environmental anthropologists, with the aim of investigating the nature and outcomes of this distinct human-primate interface. We used ethological techniques to observe the macaques' feeding behavior and activity budgets and the various ways humans interacted (or did not interact) with the macaques (Riley & Wade 2016), but we also used "thick description" (Geertz 1973) to capture, for example, "how macaques used impolite gestures from their own social worlds to generate behaviors they desired from

FIGURE 6.6 Rhesus macaques (*Macaca mulatta*) consuming provisioned food along the edge of Florida's Silver River, USA. Photo credit: Tiffany W. Wade.

humans" (Kirksey 2015: 115). We were, at times, distanced observers, and at other times, participant observers immersed in the excitable interspecies encounters that occurred along the river, and ethnographers eager to learn how different people came to regard the multispecies encounters they heard about or experienced themselves (Arluke & Sanders 1996). It was an interdisciplinary effort focused on a multispecies world. How could it have been any other way?

Lest I give the false impression that context only matters for macaques, I would like to share a fascinating non-macaque example: the human-green monkey interface in the Caribbean. Green monkeys (*Chlorocebus sabaeus*), a species native to West Africa, were introduced to the Caribbean island of St. Kitts, along with nearby Barbados and Nevis, over 300 years ago, and more recently, to the Fort Lauderdale area in Florida. Today, people-green monkey interactions in St. Kitts are increasingly common as the green monkeys venture further out of the remaining forest in search of food, including cultivated foods (Figure 6.7). In her study of St. Kitts's "monkey problem," anthropologist Kerry Dore recognized from the outset that a quantitative investigation of green monkey crop feeding behavior would be incomplete without also exploring people's perceptions of and attitudes and behavior toward the green monkeys and their foraging behavior. But, she also discovered that to understand how and why people deal (or not deal) with green

FIGURE 6.7 Adult female green monkey (*Cholorcebus sabaeus*) feeding on sugar cane on St. Kitts. Photo credit: Kerry M. Dore.

monkey crop feeding behavior, she needed to investigate the political ecology and economy of the human-environment dynamic on St. Kitts; that is, how the colonial history of St. Kitts and resulting patterns of land tenure and resource management shape how people interact with their environment, including the green monkeys. In doing so, she learned that farmers are less willing to invest in the protection of their gardens because they do not own the land, and hence, have little security about whether future cultivation of the land is even a reality. One of Dore's primary conclusions then is that "the major issue at hand in the 'monkey problem' is conflict over land ownership and associated power" (Dore 2013: 263). Dore's work exemplifies how overlapping resource between humans and other primates, and the conflict that ensues as a result, is as much, if not more, about conflicts between different groups of people (i.e., "human-human conflict," see Hill 2015) as it is about human-wildlife conflict. We will explore this issue more in Chapter 7, but in the meantime the take home point is: we (primatologists) need to be open to and learn from the broader field of anthropology, and encourage future generations of scholars to take those classes in sociocultural anthropology and environmental anthropology that they've had their eye on because they'll absolutely need them.

Anthropology and Animals

Unbeknownst to Hooton, animals have actually figured quite prominently in anthropological studies for quite some time and the rationale for their study was decidedly anthropocentric: how animals benefit humans and how animals symbolize elements of human society.[22] Anthropologist Marvin Harris (1966), for example, explained that the sacredness of Indian cattle is tied more toward their utility (e.g., production of milk and dung, animal traction) rather than adherence to the Hindu doctrine of *ahisma* (i.e., causing no harm to others). Anthropologist Mary Douglas (1957) used a symbolic approach to explain how animals figure in dietary restrictions, perceptions of fertility, and ritual practice among the Lele of then Zaire (now the Democratic of Congo).

The current wave of interest in animals in anthropology is characteristically different in a number of important ways. Contemporary human-animal inquiry has moved beyond an emphasis on traditional societies (and, for that matter, traditional anthropological themes – e.g., kinship, classification, symbols, and sustenance) toward examining the multifarious contexts in which humans and animals interface, ranging from rural- to urban- to cyber-spaces (Mullin 2002). Its practitioners have also attempted to address anthropology's blatantly anthropocentric focus by viewing animals not merely as symbols of human action and thought or as objects upon which humans act or derive economic benefit, but rather, as subjects and agents in their own right (Noske 1993; Knight 2005a). Contemporary human-animal inquiry has, in essence, shifted from an emphasis on how animals are "good to eat" (à la Marvin Harris) and "good to think" (à la Levi-Strauss) to how they are "good to live with" (Haraway 2008; Kirksey & Helmreich 2010). Scholars advancing this so-called 'animal turn' come not only from anthropology, but also from allied fields such as sociology, geography, cultural studies, and science and technology studies. Accordingly, this area of inquiry has come to be called many things, largely dependent on disciplinary parentage: anthrozoology, ethnozoology, animal studies, critical animal studies, and more recently, multispecies ethnography and multispecies anthropology. It is not my intention here to provide a historical tracing or exhaustive review of these approaches to the human-animal interface; I leave this to the multispecies experts.[23] My objective instead is to specifically demonstrate how contemporary primatology can benefit engaging with and drawing from human-animal approaches.

Seeing Individuals and Recognizing Subjectivities

I know, I know. Much like the word "reflexivity," "subjectivities" also generates a few eye rolls and may leave many primatologists feeling queasy. Bear with me though; baby steps. Seeing our study subjects as individuals is actually something that many primatologists already do. If you have spent endless hours habituating primates to your presence, then you have gotten to know them, much like hunter-gatherers get to know their prey by hunting (see Ingold 1994), so, you are half way there. If you have ever assigned your study subjects names instead of IDs or numbers, then you are even closer than half way there. But it wasn't always this way. Think, for example, of the disdain directed toward primatologist Jane Goodall when ethologists learned she broke Western scientific convention by treating the chimpanzees at Gombe as individuals: "There [Cambridge University] I was criticized for my lack of scientific method, for naming the chimpanzees rather than assigning them each a number, for *giving* them personalities, and for maintaining they had minds and emotions" (Goodall in Bekoff 2007: xxi). I specify "Western" here because individual recognition (as discussed earlier) and naming, a technique established by the late Kinji Imanishi, has long been a tradition of Japanese primatology (see Imanishi 1960).

By assigning names to primates, we are effectively affirming their individualities, giving them the opportunity to be seen as *someone* as opposed to something (Taylor 2013).[24] As sociologists Arnold Arluke and Clinton Sanders (1996) note, animals in human worlds retain a physicality to them, but they are also social constructions. In the case of primates, the process of naming transforms them from wild animal study subjects to group members whose physicalities and personalities we come to sense and know quite well. The recent "animal turn" occurring in anthropology and allied fields is particularly focused on this process because it enables us to view animals as subjects rather than objects and as parts of human worlds rather than just symbols of or adjunct to them (see Knight 2005b; Taylor 2013). The process of naming also enables researchers to develop social bonds with their study animals, which in turn positions us to be better equipped to find meaning in their behavior. Anthropologist Véronique Servais (2005) provides a particularly compelling example of the power of the social bond when retelling a story of how divers who were filmed while installing underwater equipment (to, ironically, study human-dolphin communication) failed to notice the remarkable behavior concurrently being exhibited by a male dolphin in their proximity: the dolphin imitated the divers' slow vertical fall to the sea floor as well as their position on it during the installation. Servais argues that they did not see it, and hence could not begin to appreciate the meaning behind it precisely because they did not have a social bond with the dolphin.

Once we see animals as individuals, we can then begin to appreciate their subjectivities; that is, they are not simply exhibiting behavior in relation to their environment for us to observe and record, they are experiencing that environment themselves. In turn, we are better able to understand the behavioral patterns that emerge. In their research on self-medication behavior among Kanyawara chimpanzees in Uganda, primatologist Sabrina Krief and colleagues (2006) explore the possibilities of this perspective by considering how interpreting not only what chimpanzees are doing, but also what *each* chimpanzee can interpret about their environment can help us better understand, for example, intra-community variation in the use of non-nutritional plants. Similarly, in her research on social communication among gorillas (*Gorilla gorilla*), biological anthropologist Barbara King (2004) grapples with this very issue when reflecting on the standard "checklist approach" to ethological observation that privileges the *action* of specific behavioral events to the neglect of the social meaning behind those events.[25] She cogently notes that what this approach does is fail to recognize that "[apes] are already in relation to each other even before a particular event begins…"

(p. 199). She goes on to share five social events that took place between gorillas housed at the National Zoo in Washington, DC to demonstrate that without rich description "the contingent nature of the actions" observed would have been lost (p. 203). So why would this approach to behavioral observation be useful for primatology, or biological anthropology, more broadly? King gets it: by recognizing that her study subjects can "socially construct meaning as they act together," we can come to see them as "creative communicators", which in turn, "could shift the starting point for theories into the origins of language" (p. 204).

Methodologically then, to access meaning, we need to expand our toolkit to include a sort of "narrative ethology" (Bekoff & Pierce 2009); an approach that "provides interpretation informed by [ethologists/primatologists] knowledge about a particular species and its behavior, and their attention to context and individual peculiarities" (p. 37). The stories that experienced primatologists produce are not "hard data," but they still offer considerable insight; they can provide context, expose contingency, and challenge us to ask new questions or address old ones in new ways.[26] Anthropologist Alexandra Palmer and colleagues (2015) used a narrative ethology style when relaying their ethnoprimatological research on orangutan relationships and keepers' interpretations of those relationships at Auckland Zoo, New Zealand. In addition to presenting quantitative data on dyadic interactions between orangutans, they share anecdotes from keepers, which enable us to more deeply understand how the keepers interpret orangutan social interactions and how these interpretations can then shape how they socially manage the orangutans, which in turn, can affect the nature of interactions among the orangutans. Not only does this study expand the scope of the ethnoprimatological endeavor to include urban zoological gardens – key sites where most people around the world interface with primates – but it also highlights how recognizing nonhuman subjectivities, seeing our primate study subjects as individuals with personal histories, and employing a mixed methodological approach can be of value in applied primatology situations, such as behavioral care management in captive settings. Importantly, this approach is equally valuable for applied primatology in field settings. For example, at field sites where the anthropogenic footprint is heavy, adopting such an approach enables us to see how individuals vary in how they "sense, inhabit, and experience anthropogenic spaces" (Lorimer 2010: 494), which in turn, can help us better predict how other members of their social networks (e.g., offspring) will respond. Similarly, primatologist Emmanuelle Grundmann (2006) argues that by accessing confiscated orangutans' personal histories, personalities, and links with humans we can better understand the processes of rehabilitation and reintroduction and its role in the long-term conservation of orangutans.

Primatology as Anthropology

Part of the original task bestowed to anthropological primatologists is to elucidate the general laws that govern the behavior of primates, thereby providing insight from which we can better understand the evolution of human behavior. I argue that our task should not stop there. When investigating how primates' situated ecology shapes their behavior, we encounter humans with whom primates interact, which can have implications for our interpretation of behavioral patterns. Given this reality, it is not enough to be an ethologist who solely knows a monkey as a monkey,[27] as primates who interface with humans are not clearly bounded biological subjects (Kirksey & Helmreich 2010: 556), and for that matter, neither are the humans. It is also not enough to simply study the "impact of" one on the other. Instead, we need to be able understand how humans and primates behave together. Sites where humans and primates interface do not merely involve passive observation, but rather, active engagement (Knight 2005c). We therefore need to understand *how* they interface and change one another, socially, cognitively, and physiologically (Lestel et al. 2006). We can learn, for example, from psychologist Gay Bradshaw's (2009) work exploring how "elephant breakdown" – that is, social trauma (e.g., young calves growing up without experienced mothers and allomothers) that ensues from human activities such as poaching and habitat loss – can affect elephant physiology, behavior, and culture for generations.[28] Importantly, too, the elephants' stories are also stories about us humans. Similarly, at my site in South Sulawesi where, as I described in Chapter 4, moor macaque groups are now making the road a preferred element of their home ranges. Documenting the impact of this change on their ranging patterns, diet, social interactions is just one part of the story; the story also concerns the political economic forces at play (e.g., provincial infrastructural development) and people's perceptions of wildlife and "natural spaces" that structure how they interface with the macaques. The story is decidedly multispecies.

The broadened scope that I argue for herein gives primatology an expanded purpose, one that repositions primatology as a specialization of anthropology as a whole, rather than belonging solely to the subfield of biological anthropology.[29] This is not to undermine the significance of primate studies that specifically contribute to the objectives of biological anthropology.[30] Rather, it may be time to recognize that there is not just one way of doing anthropological primatology. For many sites and for a multitude of questions there is value, and necessity, really, in placing the multispecies entanglements front and center, and to do so, we also must broaden our theoretical, epistemological, and methodological toolkits. As we have seen throughout this book, this means drawing from other approaches – including those less familiar to primatologists, such as multispecies anthropology – and integrating standard ethological techniques with ethnographic ones so that we are able to not just quantify behavior but also provide rich, qualitative description of the context and interpret the meaning that emerges from the multispecies communities that comprise humans and other primates.[31]

Following biological anthropologist Jonathan Marks (2009: 53), what is interesting about science is not "normal science" but "what happens at the conflict zones between paradigms." This is where certain threads of contemporary primatology should be situated, and recent work in ethnoprimatology shows a promising trend in this direction.[32] Anthropologists/primatologists Melissa Remis and Carolyn Jost Robinson (2017), for example, demonstrate how ecological data on primate abundance collected via line-transects in Dzanga Sangha Protected Area, Central African Republic cannot be understood without ethnographic data that reveal the history and recent trends in BaAka hunting behavior. Moreover, their multispecies approach enabled them to not only capture primate behavioral flexibility in the face of technological advances in hunting, but also how this flexibility in turn impacts BaAka men's identities as hunters and their livelihoods. They write: "incorporating multiple data sets and reflexive approaches in ethnoprimatological research is likely to better equip primatologists to talk with communities of people about forest knowledge transmission, shared vulnerabilities, human-wildlife interdependence, and ways forward to provide for a resilient future with primates in the forests upon which both depend" (p. 203).

In the next chapter, I continue this discussion of primatology *as* anthropology by exploring how an expanded theoretical and integrated methodological approach will assist us in ensuring that we still have primate populations to study in the rest of the twenty-first century and beyond.

Notes

1 In 2011, Florida's governor, Rick Scott, made disparaging remarks about Florida's universities producing students with what he perceives to be useless degrees (e.g., anthropology; see Wallace 2011). Students from the University of South Florida were quick to respond to Governor Scott by producing "This is Anthropology," an excellent presentation illustrating the multifaceted ways anthropologists make an impact in Florida and beyond (https://prezi.com/vmvomt3sj3fd/this-is-anthropology/). Making sure that anthropology's impact is seen and understood continues to be an important objective for the field. For example, the American Anthropological Association chose "Anthropology Matters" as its organizing theme for the 2017 meeting.

2 Anthroplogist Bruno Latour (1998: 231) wrote: "…what would a human be without elephants, plants, lions, cereals, oceans, ozone or plankton? Less than a human. Certainly not a human." Lestel and Taylor (2013: 183) argue though that it is not that we live with animals, but rather that "we are animals and animals are us."

3 See also Asquith (2010: 243): "significant parallels in the fieldwork process bind ethnography and primatology."

4 Habituation is often a fundamental first step in primatological fieldwork, and yet, one that is overlooked as part of the research itself. See Rees (2006), Hanson and Riley (2018).

5 Don't believe? I have witnessed eyes rolled at its mention. But see Strum and Fedigan (1999) and deWaal (2003). Moreover, this is not necessarily the case for the broader field of biological anthropology. A central theme of biological anthropologist, Jonathan Marks' (2009) *Why I am not a Scientist?*, for example, is that because "science is a human endeavor," values, interests, and politics are always at play in the production of scientific knowledge (p. 70). Similarly, in recognizing that human bodies are not just biological but "cultural-biological" – that is, our biological features are not just a product of our

genes, but also the social, ecological, and political economic environments that we live in – biological anthropologist, Alan Goodman (2013: 361) is eager to "explore the potential for a more critical and self-reflective approach to understanding human biology."

6 Reflexivity is typically viewed as part of the paradigmatic shift in sociocultural anthropology from positivist to interpretive approaches, otherwise known as the "reflexive turn" that roughly occurred in the 1980s. However, its roots are likely deeper than that (see Boyer 2015 who sees reflexivity in the Boasian tradition and in Paul Radin's "Primitive Man as Philosopher").

7 See also Despret (2008) and Candea (2010, 2013).

8 Shirley Strum and Linda Fedigan's edited volume (2000), *Primate Encounters: Models of Science, Gender, and Society*, for example, includes contributions from primatologists from a number of disciplines who address key questions such as: What is the role of gender in the science and theory building? How does science work?

9 Zoologist Thelma Rowell's work with forest-dwelling baboons was also instrumental in challenging the idea of males as central, aggressive, and dominant as a "normal" feature of baboon social life.

10 For example Fedigan (1982); Hrdy (1981); Small (1984); also see Haraway 1989 and Strum and Fedigan 1999 for extensive discussion of the role of gender in changing views of primate society.

11 See Palombit (1994), Reichard (1995), Brockelman et al. (1998), Lappan and Morino (2014).

12 For simplicity's sake, I lump North American and European into "Western" and singularize the various schools of primatology in Japan into one tradition. See Asquith (1996) and Takasaki (2000) for discussion about variability within "Japanese primatology."

13 Though Rees (2009: 222) acknowledges that an anonymous reviewer coined the term "fieldworker's regress."

14 Here I am primarily referring to the largely unchallenged claim made by primatologist Peter Rodman (1999) about the place of primatology in anthropology in his *Annual Review of Anthropology* piece titled, "Whither Primatology? Primates in Contemporary Anthropology." He wrote: "Not only is there a lack of common interest, there is a wide epistemological abyss between sociocultural anthropology and primatology" (p. 331).

15 I make this latter point because sometimes the anthropogenic footprint is not immediately (temporally) visible, as the fieldwork story that I begin Chapter 5 with demonstrates.

16 Up until fairly recently, the Lindu plain was only reachable by foot or donkey via a 17-km trail through the forest that connects it to the primary provincial road leading to the provincial capital of Palu. In 2003, once the tilapia fishing industry exploded, the trail was widened to better accommodate motorcycle travel and transport of fish to the Palu valley. In 2012, following a 6.2 magnitude earthquake that struck the plain, the trail was further widened to allow vehicles to transport emergency services and supplies, but rain and instability of the trail make it frequently impassable by anything but motorcycles and people on foot.

17 At the time I lived there (2002–2004), Tomado was inhabited by approximately 132 households, 90 percent of which were *To Lindu* and the remainder were Bugis people from South Sulawesior other Kaili people from Kulawi.

18 Power et al. (2006: 609) define a keystone species as "a species whose effect is large, and disproportionally large relative to its abundance."

19 For example, rhesus macaques (*Macaca mulatta*) have extended their range into peninsular India; originally the land of the bonnet macaque (*Macaca radiata*) (Kumar et al. 2011).

20 Provisioning is the deliberate offering of food to animals. At the Silver River, two feeding sites were established on either side of the river where commercial monkey chow and fruit were provided twice daily (Sarris 1980).

21 Between 1986–1990, the University of Florida Veterinary College sterilized 20 females by tubal ligation (Montague et al. 1994). The trapping program resulted in the removal of 500 monkeys from the banks of the Silver River between 1984–1993 (Wolfe 2002).

22 Here I refer to his noting that "...the anthropologist rarely, if ever, studies any other animal than man." (Hooton 1942: xxxi). Although, to be fair, Hooton was already deceased (in 1954) when many of the early anthropological works on humans and animals emerged. See Shanklin (1985) for a review of anthropology's interest in the function and meaning of animals.

23 See Shanklin (1985), Mullin (1999), Arluke (2003), Haraway (2008), Kirksey and Helmreich (2010), Parreñas (2012), Ogden et al. (2013), Hurn (2017).

24 Although as rightly noted by one of Amanda Rees' primatologist interviewees, "once you call an animal 33, that's his name, it becomes his name, his name is 33 as if you'd call him May or June" (Rees 2001: 238). Mary Phillips (1994) reveals this in the opposite by exploring the meaning behind <u>not</u> naming animals at biomedical facilities. She argues that a proper name provides an individual a social identity, a biography of sorts. In laboratory settings then, by not doing so, researchers generate the social construction of the "laboratory animal," which is categorically distinct from the pet animal who is loved and nurtured. The lack of naming also serve to emotionally distance researchers and technicians from animals whose tenure at the labs may end in death.

25 Psychologist William Hopkins (2005), in his published review, is critical of Barbara King's 2004 book titled *The Dynamic Dance: Nonvocal Communication in the African Great Apes* for "focusing largely on descriptive data" and relying "heavily on anecdotal reports on social communication in the African great apes," which of course is precisely what King, in her 2004 article in *Social Anthropology*, and myself here, are trying to argue *for*.

26 By incorporating narrative ethology into our overall research approach we would essentially be blending fundamental elements of the Japanese tradition of primatology, such as recognizing value in anecdotes and descriptive data about individuals, with the Western tradition of primatology.

27 Or an ape as an ape or a lemur as a lemur, etc. You get the idea. See Lestel et al. 2006: 170.

28 See also Bradshaw et al. 2005.

29 Incidentally, Mullin (2002: 387) lists primatology among other specializations of the overall field of anthropology: "Beyond these basic four fields, one could further divide anthropology into a nearly endless array of specializations (primatology, legal anthropology, medical anthropology, historical archaeology, to name just a few)." Also, by doing so we would see another way the Western tradition of primatology converges on the Japanese tradition, specifically the Kyoto school, as Takasaki (2000: 160) notes "many, if not all, of the Kyoto school fieldworkers regard themselves primarily as anthropologists not primatologists."

30 Although, I would argue that all primatological studies should be minimally aware of the anthropogenic context within which they are likely embedded – as this is part of

conducting ethically informed field primatology (Mackinnon & Riley 2010; Riley & Bezanson 2018; Code of Best Practices for Field Primatology; www.asp.org/resources/docs/Code%20of_Best_Practices%20Oct%202014.pdf).

31 It is important to point out that this learning from others is not uni-directional; ethnoprimatology has served as a model for emergence of an interdisciplinary research program focused on human-elephant entanglements, *ethnoelephantology* (Locke 2013).

32 Also see the special issue "Ethnoprimatology in the 21st Century" (*International Journal of Primatology*, October 2018, Volume 39, Issue 5,), in which several of the papers illustrate the value of applying multispecies approaches to the study of the human-primate interface.

References

Acciaioli GL. 1989. *Searching for Good Fortune: The Making of a Bugis Shore Community at Lake Lindu, Central Sulawesi*. PhD Thesis. Australian National University.

Altmann J. 1974. Observational study of behavior: Sampling methods. *Behavior* 49: 227–67.

Arluke A. 2003. Ethnozoology and the future of sociology. *International Journal of Sociology and Social Policy* 23: 26–45.

Arluke A, Sanders CR. 1996. *Regarding Animals*. Philadelphia: Temple University Press.

Asquith PJ. 1991. Primate research groups in Japan: Orientations and East-West differences. In *The Monkeys of Arashiyama: Thirty-five Years of Research in Japan and the West*, ed. L Fedigan, PJ Asquith, pp. 81–98. New York: SUNY Press.

Asquith PJ. 1996. Japanese science and Western hegemonies: Primatology and the limits to set questions. In *Naked Science: Anthropological Inquiry into Boundaries, Power, and Knowledge*, ed. L Nader, pp. 239–56. New York: Routledge.

Asquith PJ. 2000. Negotiating science: Internationalization and Japanese primatology. In *Primate Encounters: Models of Science, Gender, and Society*, ed. SC Strum, L Fedigan, pp. 165–83. Chicago: Chicago University Press.

Baynes-Rock M. 2013. Life and death in the multispecies commons. *Social Science Information* 52: 210–27.

Bekoff M. 2007. *The Emotional Lives of Animals. A Leading Scientist Explores Animal Joy, Sorrow, and Empathy – And Why They Matter*. Novato: New World Library.

Bekoff M, Pierce J. 2009. *Wild Justice: The Moral Lives of Animals*. Chicago: University of Chicago Press.

Boyer D. 2015. Reflexivity reloaded: From anthropology of intellectuals to critique of method to studying sideways. In *Anthropology Now and Next: Essays in Honor of Ulf Hannerz*, ed. TH Eriksen, G Garsten, S Randeria, pp. 91–110. New York: Berghahn.

Bradshaw GA. 2009. *Elephants on the Edge: What Animals Teach Us About Humanity*. Yale University Press.

Bradshaw GA, Schore AN, Brown JL, Poole JH, Moss CJ. 2005. Elephant breakdown. *Nature* 433: 807.

Brockelman WY, Reichard U, Treesucon U, Raemaekers JJ. 1998. Dispersal, pair formation and social structure in gibbons (*Hylobates lar*). *Behavioral Ecology and Sociobiology* 42: 329–39.

Candea M. 2010. "I fell in love with Carlos the meerkat": Engagement and detachment in human-animal relations. *American Anthropologist* 37: 241–58.

Candea M. 2013. Habituating meerkats and redescribing animal behaviour science. *Theory, Culture & Society* 30: 105–28.

Cerroni-Long EL. 1996. Human science. *Anthropology News* January: 50–52.

Davis G. 1976. *Parigi: A Social History of the Balinese Movement to Central Sulawesi, 1907–1974*. Stanford University.

Davis TA. 1988. Uses of semi-wild palms in Indonesia and elsewhere in South and Southeast Asia. *Advances in Economic Botany* 6: 98–118.

de Waal FB. 2003. Silent invasion: Imanishi's primatology and cultural bias in science. *Animal Cognition* 6: 293–99.

de Waal FBM. 1989. *Peacemaking Among Primates*. Cambridge, MA: Harvard University Press.

Despret V. 2008. Culture and gender do not dissolve into how scientists "read" nature: Thelma Rowell's heterodoxy. In *Rebels, Mavericks, and Heretics in Biology*, ed. O Harman, M Dietrich, pp. 338–55. New Haven, CT: Yale University Press.

DeVore I. 1965. Male dominance and mating behavior in baboons. In *Sex and Behavior*, ed. FA Beach, pp. 266–89. New York: John Wiley & Sons.

Dore KM. 2013. *An Anthropological Investigation of the Dynamic Human-Vervet Monkey (Chlorocebus aethiops sabaeus) Interface in St. Kitts, West Indies*. The University of Wisconsin-Milwaukee.

Dore KM. 2017. Ethnophoresy. In *The International Encyclopedia of Primatology*. John Wiley & Sons, Inc.

Douglas M. 1957. Animals in Lele religious symbolism. *Africa* 27: 46–58.

Fedigan LM. 1982. *Primate Paradigms: Sex Role and Social Bonds*. Montreal: Eden Press.

Fuentes A. 2000. Hylobatid communities: changing views on pair bonding and social organization in hominoids. *Yearbook of Physical Anthropology* 43: 33–60.

Fuentes A. 2010a. Naturalcultural encounters in Bali: Monkeys, temples, tourists, and ethnoprimatology. *Cultural Anthropology* 25: 600–24.

Fuentes A. 2010b. Measuring meaning and understanding in primatological and biological anthropology fieldwork: context and practice. In *Centralizing Fieldwork: Critical Perspectives from Biological and Social Anthropology*, ed. J MacClancy, A Fuentes, pp. 121–36. New York: Berghan Books.

Fuentes A, Southern M, Suaryana KG. 2005. Monkey forests and human landscapes: Is extensive sympatry sustainable for *Homo sapiens* and *Macaca fascicularis* on Bali? In *Commensalism and Conflict: The Primate-Human Interface*, ed. J Paterson, J Wallis, pp. 168–95. Norman, OK: American Society of Primatologists.

Geertz C. 1973. *The Interpretation of Cultures: Selected Essays*. New York: Basic Books, Inc.

Goodman AH. 2013. Bringing culture into human biology and biology back into anthropology. *American Anthropologist* 115: 359–73.

Grundman E. 2006. Back to the wild: Will reintroduction and rehabilitation help the long-term conservatin of orangutans in Indonesia? *Social Science Information* 42: 265–84.

Gumert MD. 2011. The common monkey of Southeast Asia: Long-tailed macaque populations, ethnophoresy, and their occurrence in human environments. In *Monkeys on the Edge: Ecology and Management of Long-tailed macaques and their Interface with Humans*, ed. MD Gumert, A Fuentes, L Jones-Engel, pp. 3–44. Cambridge: Cambridge University Press.

Hanson KT, Riley EP. 2018. Beyond neutrality: The human–primate interface during the habituation process. *International Journal of Primatology* 39: 852–77.

Haraway DJ. 1989. *Primate Visions: Gender, Race, and Nature in the World of Modern Science*. New York: Routledge.

Haraway DJ. 2003. *The Companion Species Manifesto: Dogs, People, and Significant Otherness*. Chicago: Prickly Paradigm Press.

Haraway DJ. 2008. *When Species Meet*. Minneapolis: University of Minnesota Press.

Harris M. 1966. The cultural ecology of India's sacred cattle. *Current Anthropology* 7: 51–66.

Heinsohn T. 2003. Animal translocation: long-term human influences on the vertebrate zoogeography of Australasia (natural dispersal versus ethnophoresy). *Australian Zoologist* 32: 351–76.

Hill CM. 2015. Perspectives of "Conflict" at the wildlife–agriculture boundary: 10 years on. *Human Dimensions of Wildlife* 20: 296–301.

Hill DA, van Hooff JARAM. 1994. Affiliative relationships between males in groups of nonhuman primate: A summary. *Behaviour* 130: 143–49.

Hooton EA. 1942. *Man's Poor Relations*. Doubleday, Doran.

Hopkins WD. 2005. Book review: The dynamic dance: Nonvocal communication in African great apes. *International Journal of Primatology* 26: 1479–81.

Hrdy SB. 1981. *The Woman that Never Evolved*. Cambridge, MA: Harvard University Press.

Hurn S. 2017. Human-animal relations (HAR). In *International Encyclopedia of Primatology*, pp. 1–7: John Wiley & Sons, Inc.

Hymes D. 1972. *Reinventing Anthropology*. New York: Pantheon Books.

Imanishi K. 1960. Social organization of subhuman primates in their natural habitat. *Current Anthropology* 1: 393–407.

Ingold T. 1994. From trust to domination: An alternative history of human-animal relations. In *Animals and Human Society: Changing Perspectives*, ed. A Manning, J Serpell, pp. 1–22. London: Taylor & Francis.

Itani J. 1961. The society of Japanese monkeys. *Japan Quarterley* 8: 421–30.

King BJ. 2004. Towards an ethnography of African great apes. *Social Anthropology* 12: 195–207.

Kinnaird MF, O'Brien TG. 2005. Fast foods of the forest: The influence of figs on primates and hornbills across Wallace's line. In *Tropical Fruits and Frugivores*, ed. JL Dew, JP Boubli, pp. 155–84. Dordrecht, The Netherlands: Springer.

Kirksey E. 2015. *Emergent Ecologies*. Durham: Duke University Press.

Kirksey SE, Helmreich S. 2010. The emergence of multispecies ethnography. *Cultural Anthropology* 25: 545–76.

Knight J. 2005a. Introduction. In *Animals in Person: Cultural Perspectives on Human-Animal Intimacy*, ed. J Knight, pp. 1–13. Oxford: Berg.

Knight J, ed. 2005b. *Animals in Person: Cultural Perspectives on Human-Animal Intimacy*. Oxford: Berg.

Knight J. 2005c. Feeding Mr. Monkey: Cross-species food "exchange" in Japanese Monkey Parks. In *Animals in Person: Cultural Perspectives on Human-Animal Intimacy*, ed. J Knight, pp. 231–53. Oxford: Berg.

Knight J. 2017. Introduction to Part II. In *Ethnoprimatology: A practical guide to research at the human-nonhuman primate interface*, ed. KM Dore, EP Riley, A Fuentes, pp. 171–75. Cambridge: Cambridge University Press.

Krief S, Wrangham R, Lestel D. 2006. Diversity of items of low nutritional value ingested by chimpanzees from Kanyawara, Kibale National park, Uganda: An example of etho-ethnology of chimpanzees. *Social Science Information* 45: 227–63.

Kumar R, Radhakrishna S, Sinha A. 2011. Of least concern? Range extension by rhesus macaques (*Macaca mulatta*) threatens long-term survival of bonnet macaques (*M. radiata*) in peninsular India. *International Journal of Primatology* 32: 945–59.

Lane KE, Holley C, Hollocher H, Fuentes A. 2011. The anthropogenic environment lessens the intensity and prevalence of gastrointestinal parasites in Balinese long-tailed macaques (*Macaca fascicularis*). *Primates* 52: 117–28.

Lane KE, Lute M, Rompis A, Wandia IN, Arta Putra IGA, et al. 2010. Pests, peestilence, and people: The long-tailed macaque and its role in the cultural complexities of Bali. In *Indonesian Primates*, ed. S Gursky-Doyen, J Supriatna, pp. 235–48. New York: Springer.

Lappan S, Morino L. 2014. Mating in the presence of a competitor: Audience effects may promote male social tolerance in polyandrous siamang (*Symphalangus syndactylus*) groups. *Behaviour* 151: 1067–89.

Latour B. 1998. To modernize or to ecologize? That is the question. In *Remaking Reality: Nature at the Millenium*, ed. B Braun, N Castree, pp. 221–43. Routledge: London.

Latour B, Woolgar S. 2013. *Laboratory Life: The Construction of Scientific Facts*. Princeton: Princeton University Press.

Lestel D, Brunois F, Gaunet F. 2006. Etho-ethnology and ethno-ethology. *Social Science Information* 45: 155–77.

Lestel D, Taylor H. 2013. Shared life: An introduction. *Social Science Information* 52: 183–86.

Li TM. 2000. Articulating indigenous identity in Indonesia: Resource politics and the tribal slot. *Comparative Studies in Society and History* 42: 149–79.

Lichterman P. 2017. Interpretive reflexivity in ethnography. *Ethnography* 18: 35–45.

Locke P. 2013. Explorations in ethnoelephantology: Social, historical, and ecological intersections between Asian elephants and humans. *Environment and Society: Advances in Research* 4: 79–97.

Lorimer J. 2010. Elephants as companion species: The lively biogeographies of Asian elephant conservation in Sri Lanka. *Transactions of the Institute of British Geographers* 35: 491–506.

MacKinnon K, Riley EP. 2010. Field primatology of today: Current ethical issues. *American Journal of Primatology* 72: 749–53.

Madden R. 2010. *Being Ethnographic: A Guide to the Theory and Practice of Ethnography*. London: Sage.

Maples WR, Brown AB, Hitchens PM. 1976. Introduced monkey populations at Silver Springs, Florida. *The Florida Anthropologist* 29: 133–36.

Marks J. 2009. *Why I Am Not a Scientist: Anthropology and Modern Knowledge*. Berkeley, CA: University of California Press.

Mogea J, Seibert B, Smits W. 1991. Multipurpose palms: the sugar palm (*Arenga pinnata* (Wurmb) Merr.). *Agroforestry Systems* 13: 111–29.

Montague CL, Colwell SV, Percival HF, Gottgens JF. 1994. Issues and options related to the management of Silver Springs rhesus macaques, Technical Report #49. Completed for the Florida Game and Fresh Water Fish Commission.

Mullin MH. 2002. Animals and anthropology. *Society and Animals* 10: 387–94.

Mullin MH. 1999. Mirrors and windows: Sociocultural studies of human-animal relationships. *Annual Review of Anthropology* 28: 201–24.

Nader L. 2001. Anthropology! Distinguished Lecture–2000. *American Anthropologist* 103: 609–20.

Noske B. 1993. The animal question in anthropology: A commentary. *Society and Animals* 1: 185–90.

Ogden LA, Hall B, Tanita K. 2013. Animals, plants, people, and things: A review of multispecies ethnography. *Environment and Society* 4: 5–24.

Ostrom E. 1999. Self-governance and forest resources. *CIFOR Occasional Papers*: 1–14.

Palmer A, Malone N, Park J. 2015. Accessing orangutans' perspectives: Interdisciplinary methods at the human/animal interface. *Current Anthropology* 56: 571–78.

Palombit R. 1994. Dynamic pair bonds in Hylobatids: Implications regarding monogamous social systems *Behaviour* 128: 65–101.

Parreñas RJS. 2012. Producing affect: Transnational volunteerism in a Malaysian orangutan rehabilitation center. *American Ethnologist* 39: 673–87.
Phillips MT. 1994. Proper names and the social construction of biography: The negative case of laboratory animals. *Qualitative Sociology* 17: 119–42.
Power ME, Tilman D, Estes JA, Menge BA, Bond WJ, et al. 1996. Challenges in the quest for keystones. *BioScience* 46: 609–20.
Rees A. 2001. Anthropomorphism, anthropocentrism, and anecdote: primatologists on primatology. *Science, Technology, & Human Values* 26: 227–47.
Rees A. 2006. A place that answers questions: primatological field sites and the making of authentic observations. *Studies in History and Philosophy of Biological and Biomedical Sciences* 37: 311–33.
Rees A. 2009. *The Infanticide Controversy: Primatology and the Art of Field Science*. Chicago: University of Chicago Press.
Reichard U. 1995. Extra-pair copulations in a monogamous gibbon (*Hylobates lar*). *Ethology* 100: 99–112.
Remis MJ, Jost Robinson CA. 2017. Nonhuman primates and "others" in the Dzanga Sangha Reserve: The role of anthropology and multispecies approaches in ethnoprimatology. In *Ethnoprimatology: A Practical Guide to Research at the Human-Primate Interface*, ed. KM Dore, EP Riley, A Fuentes, pp. 190–205. Cambridge: Cambridge University Press.
Reynolds V. 1992. The discovery of primate kinship systems by Japanese anthropologists. In *Topics in Primatology 2*, ed. N Itoigawa, Y Sugiyama, G Sackett, R Thompson, pp. 79–87. Tokyo: Univerisity of Tokyo Press.
Riley EP. 2013. The human–macaque interface in the Sulawesi Highlands. In *The Politics of Species: Reshaping our Relationships with Other Animals*, ed. R Corbey, A Lanjouw, pp. 189–96. Cambridge: Cambridge University Press.
Riley EP, Bezanson M. 2018. Ethics of primate fieldwork: Toward an ethically-engaged primatology. *Annual Review of Anthropology* 47: 493–512.
Riley EP. 2007a. The human-macaque interface: Conservation implications of current and future overlap and conflict in Lore Lindu National park, Sulawesi, Indonesia. *American Anthropologist* 109: 473–84.
Riley EP. 2007b. Flexibility in diet and activity patterns of *Macaca tonkeana* in response to anthropogenic habitat alteration. *International Journal of Primatology* 28: 107–33.
Riley EP. 2010. The importance of human-macaque folklore for conservation in Lore Lindu National Park, Sulawesi, Indonesia. *Oryx* 44: 235–40.
Riley EP, Wade TW. 2016. Adapting to Florida's riverine woodlands: the population status and feeding ecology of the Silver River rhesus macaques and their interface with humans. *Primates* 57: 195–210.
Rodman PS. 1999. Whither primatology? The place of primates in contemporary anthro pology. *Annual Review of Anthropology* 28: 311–39.
Rowell T. 2000. A few peculiar primates. In *Primate Encounters*, ed. SC Strum, L Fedigan, pp. 57–70. Chicago: University of Chicago Press.
Sarris E. 1980. Aspects and implications of supplemental foraging among provisional monkeys. *Florida Scientist* 43: 164–68.
Scholte B. 1972. Toward a reflexive and critical anthropology. In *Reinventing Anthropology*, ed. D Hymes, pp. 430–57. New York: Pantheon Books.
Servais V. 2005. Enchanting dolphins: An analysis of human-dolphin encounters. In *Animals in Person: Cultural Perspectives on Human-Animal Intimacies*, ed. J Knight, pp. 211–30. Oxford: Berg.
Shanklin E. 1985. Sustenance and symbol: Anthropological studies of domesticated animals. *Annual Review of Anthropology* 14: 375–403.

Small MF. 1984. *Female Primates: Studies by Women Primatologists.* New York: Alan R. Liss.

Strier KB. 1994. Myth of the typical primate. *Yearbook of Physical Anthropology* 37: 233–71.

Strier KB. 2003. Primate behavioral ecology: from ethnography to ethology and back. *American Anthropologist* 105: 16–27.

Strier KB. 2011. Why anthropology needs primatology. *General Anthropology* 18: 6–8.

Strum SC. 1982. Agonistic dominance in male baboons: An alternative view. *International Journal of Primatology* 3: 175.

Strum SC. 1987a. *Almost Human.* Chicago: University of Chicago Press.

Strum SC. 1987b. The "gang" moves to strange new land. *National Geographic* 172: 676–90.

Strum SC, Fedigan L. 1999. Theory, method, gender, and culture: What changed our views of primate society? In *The New Physical Anthropology: Science, Humanism, and Critical Reflection*, ed. SC Strum, DG Lindburg, D Hamburg, pp. 67–105. Upper Saddle River, NJ: Prentice Hall.

Strum SC, Fedigan LM. 2000. *Primate Encounters: Models of Science, Gender, and Society.* Chicago: University of Chicago Press.

Sussman RW, Garber PA, Cheverud JM. 2005. The importance of cooperation and affiliation in the evolution of primate sociality. *American Journal of Physical Anthropology* 128: 84–97.

Takasaki H. 2000. Traditions of the Kyoto School of field primatology in Japan. In *Primate Encounters: Models of Science, Gender, and Society*, ed. SC Strum, L Fedigan, pp. 151–64. Chicago: Chicago University Press.

Taylor N. 2013. *Humans, Animals, and Society: An Introduction to Human-Animal Studies.* Brooklyn, NY: Lantern Books.

Tokuyama N, Furuichi T. 2016. Do friends help each other? Patterns of female coalition formation in wild bonobos at Wamba. *Animal Behaviour* 119: 27–35.

Wallace J. 2011. Rick Scott no fan of anthropology. In *Sarasota Herald-Tribune,* Monday, October 10, 2011; http://politics.heraldtribune.com/2011/10/10/rick-scott-no-fan-of-anthropology/

Wheatley B. 1999. *The Sacred Monkeys of Bali.* Prospect Heights, IL: Waveland Press, Inc.

Wolfe LD. 2002. Rhesus macaques: A comparative study of two sites, Jaipur, India, and Silver Springs, Florida. In *Primates Face to Face: The Conservation Implications of Human-Nonhuman Primate Interconnections*, ed. A Fuentes, LD Wolfe, pp. 310–30. Cambridge: Cambridge University Press.

7

PRIMATE CONSERVATION IN THE TWENTY-FIRST CENTURY AND BEYOND

When field primatology emerged in earnest in the 1950s–1960s, scholars focused directly on their topic of interest – social behavior, diet and foraging behavior, locomotor patterns – without worrying too much about whether their study populations would decline or even perish, for that matter. But, by the 1970s, primatologists began paying attention. In 1978, primatologist and conservationist Russell Mittermeier published the first version of *A Global Strategy for Primate Conservation*, which provides an action program to achieve the International Union for the Conservation of Nature (IUCN) Primate Specialist Group's goals of ensuring the survivability of threatened primates and protecting them in areas in high diversity and abundance.[1] Indeed, things were a bit different in that era: the human population size was only 3.9 billion (compared to the 2017 estimate of 7.55 billion) and the extent of forested habitat was much greater.[2] Fast-forward three decades, and the call for conservation action indicates a much more dire need. For example, behavioral ecologist Charles Janson (2000) wrote: "It does not seem far-fetched to claim that virtually all primate behavioral ecologists in the future will have to be conservation activists if they wish to have anything to study at all" (2000:74). Nope, not far-fetched at all, for that future has come: recent estimates indicate that approximately 60 percent of primates are threatened with extinction, with the primary causal factor being unsustainable human practices (Estrada et al. 2017). While, globally, the primary threat is the "expanding agricultural frontier," secondary threats, which vary by region, include infrastructural development (e.g., hydro-electric projects, road networks), hunting and trade, forest fragmentation and degradation due to activities such as logging, mining, cattle ranching, and climate change (Estrada et al. 2017; Figure 7.1). Moreover, the shrinking and loss of primate habitats result in increased spatial proximity with humans, and consequently, greater potential for disease transmission, with important implications for

FIGURE 7.1 Clearing of trees for agricultural development within the home range of a group of Tonkean macaques (*Macaca tonkeana*), Lindu, Central Sulawesi, Indonesia. Photo credit: Erin P. Riley.

primate survivability and public health (Bublitz et al. 2015; Calvignac-Spencer et al. 2012). In this chapter, I cover the remaining component of the promise of contemporary primatology: actively engaging in the conservation of primates and their habitats, and doing so in an integrative way by drawing from the natural sciences, the social sciences, and the humanities.

With more than half of all primates being threatened with extinction, a predicted human population size of more than 9 billion people by 2050, a net loss of 7 million hectares of forest per year in tropical countries in just the first decade of the twenty-first century, primate conservation is undoubtedly a formidable challenge at best.[3] It is not surprising then that primatologists may be reticent about becoming involved. Conservation work is challenging and often depressing. To compound this, there are other factors that likely dissuade primatologists (and others) from becoming involved. These factors include: the concern that conservation-oriented work is less prestigious (read, mostly descriptive and theory-light) and hence isn't valued (e.g., for tenure and promotion, for publication in high-impact journals) the same way as more theoretically-oriented work;[4] the notion that conservation is the domain of wildlife managers, non-governmental organizations (NGOs), and government agencies not individual researchers; the concern that primatologists, particularly those doing behavioral ecology who are interested in individual variation in behavior, work at different scales than conservation practitioners who are concerned with population and community-level dynamics; the goal of producing scholarly publications takes precedence over engaging in conservation action; or that scholars hold the perception that their domain of study (e.g., evolution) is not relevant to conservation.[5]

Despite these impediments, the promising news is that it appears that more and more primatologists consider conservation-oriented questions and topics to be an important part of the future of primatology. In 2013, Joanna Setchell, biological anthropologist and Editor-in-Chief of the *International Journal of Primatology*, published the results of a survey she conducted on the "Top 10 Questions in Primatology" administered to the membership of the International Primatological Society (IPS), which at the time comprised 1430 members.[6] Thirty-two people responded to the survey and identified 170 questions that are important to the future of primatology. Conservation was the most frequent theme, constituting ¼ of the 170 questions (Setchell 2013). Although this survey had a low response rate, it is indicative that a portion of the IPS membership is thinking about a number of conservation concerns, including identifying the best ways to conserve primates, how primates will adapt to habitat change, how to change attitudes toward primates, how our research practices may negatively impact conservation, how climate change and other anthropogenic processes affect primate persistence, and how people and primates can coexist. These results suggest that at least some primatologists have an interest in conservation, and are paying greater attention to the role humans and human behavior play in shaping primate behavior as well as the habitats primates need to survive. And, importantly, this increased attention does not merely reside in primatologists' minds and mouths, but also in the work they do. For example, using the Web of Science™ database, Matthew McLennan and colleagues (2017) found a notable growth in the amount of literature on primates in anthropogenic spaces since the 1970s, with the greatest percentage of publications coming from the current decade (2010s+). So, we are witnessing a trend toward greater attention to conservation issues among primatologists; excellent! Now, let's see what we can learn from conservation scholars, from both the natural and social sciences, to advance a contemporary primatology that is theoretically-driven, conservation-oriented, and ethically-engaged.

Behavioral Knowledge Matters for Conservation

Remember how during its adolescent years primatology borrowed from other fields, such as ethology and behavioral ecology, for theoretical and methodological guidance? Well, it is with the question of "how behavior can contribute to conservation?" that we can rely on these fields yet again for guidance. In the 1990s, when ethologists and behavioral ecologists increasingly saw that their study populations were disappearing, they began thinking more seriously about how their knowledge of behavior could be incorporated into conservation management. This flurry of interest resulted in a number of publications, including a series of review papers and edited volumes, and the formation of a conservation committee for the Animal Behavior Society (Caro 2007; Angeloni et al. 2008). Part of the rationale was that in order for the then nascent "crisis discipline" of conservation biology (Soulé 1985) to be effective, it needed a diverse toolkit. While that toolkit initially included expertise from expected fields such as population genetics, ecology, physiology, public policy, to name a few, behavioral ecology came to be seen as an important addition. How could behavioral data be ignored given the important role it plays as mediator between an animal's fitness and its environment? Enter 'conservation behavior' – a research area that investigates "how proximate and ultimate aspects of the behavior of an animal can be of value in preventing the loss of biodiversity" (Bucholtz 2007: 401). Within this framework, the primary fitness-related behavior that behavioral ecologists, including primate behavioral ecologists, study – foraging, ranging and habitat selection, predation avoidance, social and reproductive behavior – have not only theoretical significance, but also applied relevance.

The conservation behavior framework is particularly well-suited for the realities of the twenty-first century. As anthropogenic activities continue to alter the environments in which animals live, understanding how they are able to behaviorally respond to these impacts is a critical component for designing conservation management efforts. Understanding which species are able to move between fragments and how they do so, for example, will help us determine which will survive and which will not. For example, at Kibale National Park in Uganda, although red tailed monkeys (*Cercopithecus ascanius*) readily move between fragments using corridors and unforested areas as passageways, blue monkeys, a closely related species with a similar diet and social organization as red tailed monkeys, do not (Chapman & Peres 2001). However, at another site in Uganda, Budongo Forest Reserve, blue monkeys in logged forests had smaller home ranges, used specific areas of the home range more intensely, but exhibited longer daily path lengths than those in unlogged forests, which suggests they had to travel further in a day to access fruit patches (Fairgrieve 1995). This example hints at the importance of considering the diversity of behavioral strategies that result from plasticity in response to anthropogenic habitat alteration. And, importantly, failure to account for this diversity could lead to misinformed management and conservation efforts. For example, in their study on brown bear (*Ursus arctos*) behavioral flexibility in human-altered landscapes in Turkey, ecologist Gabriele Cozzi and colleagues (2016) documented two behavioral types within the bear population: those that exploited a city garbage dump and those that did not, with concomitant differences in migration patterns and habitat selection. The authors conclude (p.8):

> Therefore, identification and consideration of observed variation in behavioral types is fundamental for the correct implementation of evidence-based

> conservation strategies. Failures to detect such differences could result in the erroneous allocation of limited conservation resources, such as setting aside portions of land characterized by landscape features that are critical to only particular behavioral types.

This growth of interest in applying behavioral knowledge to address conservation problems in the last 30 years has, however, also been paired with considerable self-reflection on the true extent of integration between behavioral ecology and conservation biology.[7] In other words, has the core research that derives from behavioral ecological studies actually made practical contributions to conservation action? Conservation biologist Tim Caro's (2007) primary concern is that while descriptive behavioral data on, for example, how animals adjust their ranging behavior and habitat use in response to human activities, has been useful, theoretically-driven behavioral ecology research has yet to be applied.[8] For example, knowledge about mating systems was considered particularly relevant for conservation because of its influence on effective population size (N_e) – a measure of a population's ability to maintain genetic diversity. However, "knowing the outcome of these effects [of different mating systems] does not suggest immediate management action, as it is difficult to alter the mating system of a species" (Caro 2007: 394). Yet, mating and breeding systems are just one area of focus of behavioral ecology. Foraging behavior, a major domain of behavior with important fitness consequences, and one guided by considerable theory, is particularly relevant for conservation. For example, examining how animals flexibly adjust (or not adjust) their foraging behavior in response to anthropogenic habitat alteration can simultaneously be placed in a behavioral ecological framework *and* have conservation applications. As we learned in Chapter 4, the anthropogenic environments in which primates live today may differ from the environments in which they evolved in terms of the specific environmental details, but not necessarily in terms of the trade-offs they present. We can therefore use guiding theoretical frameworks, such as optimal foraging theory, to make predictions about how we expect primates to respond to those trade-offs.

Feeding on agricultural crops by primates and other wildlife, a behavior we explored in Chapters 4 and 5, is an excellent example of a type of foraging behavior that has multiple fitness consequences. Agricultural land can provide spatially-concentrated foods that may be easy to harvest and have important nutritional benefits. At the same, crop feeding is a risky behavior because it could be met with retaliation by farmers. To better understand how primates deal with these trade-offs when crop feeding, primatologist Sabrina Krief and colleagues (2014) examined crop feeding behavior by chimpanzees (*Pan troglydytes*) in Kibale National park, Uganda, predicting that they would use characteristic behavioral strategies (e.g., vigilance, modifying party size) to maximize benefits (i.e., consume the crops but avoid being detected by farmers). Using camera trap technology to indirectly observe crop feeding behavior, they found, among other things, that the chimpanzees extended their feeding excursions into well after sun down – a time that chimpanzees are typically resting in their nests. This study not only illuminates behavioral adjustments made by chimpanzees in anthropogenic environments, but it also provides behavioral data that could be incorporated into a mitigation plan for crop feeding behavior. For example, farmers and park authorities could begin rotating teams responsible for extending guarding hours into the nighttime hours.

The study of flexible foraging behavior, such as crop feeding, and its conservation management implications can also be guided by life history theory. Life history theory predicts that differences in fitness expectations should lead to differences in risk taking behavior, such that those individuals who have more to lose in terms of future reproductive potential should be more risk-averse compared to individuals who have a high current reproductive potential (Wolf et al. 2007). Access to human food may represent an important source of energy for individuals at specific life history markers, such as sexual maturation, a period associated with a peak in energy requirements. For example, the finding that adult chacma baboons (*Papio hamadryas ursinus*), including lactating females, initiated more crop feeding events near Hwange National Park, Zimbabwe than other age classes suggests that energetic demands may drive this behavior (Schweitzer et al. 2017). Accordingly, the authors conclude that mitigation efforts should focus on making crops less palatable to the baboons. Similarly, biologist Patrick Chiyo and colleagues (2012), in their study examining factors contributing to the feeding on agricultural crops by male elephants (*Loxodonta africana*) in Kenya, found that the odds of being a crop feeder increased with age and were doubled for males at their reproductive peak. These results suggest that reproduction poses energetic costs for males that may be offset by access to human food. At the same time, this study found that social learning plays an important role in the spread of this behavior across different age classes, with younger males learning from older males. These behavioral data, derived from theoretically-driven research, can in turn inform the management of human-elephant conflict. For example, the authors suggest that adverse conditioning techniques, such as pepper spray canisters, could be targeted toward older males to minimize social transmission of the behavior, but also toward younger males so that they are less likely to crop feed when they reach their reproductive peak.

These examples demonstrate that theoretically-driven behavioral ecology research has much to contribute to conservation management concerns. These examples also exemplify how researchers whose primary interest may be in conservation outcomes can devise theoretically-driven research agenda that will inform those outcomes. This is an important distinction, and one that I hope will dispel the mistaken assumption that conservation-oriented research is atheoretical.

Rethinking Conservation: Conservation of What, for Whom?

The traditional view of conservation holds that the primary objective is to preserve biological diversity. Ecologists have amassed copious datasets that point to the critical role biodiversity plays in ecosystem function and for ecological services, so there is no doubt about the importance of its preservation.[9] What lies at the heart of this conventional view though is the postulate that nature has intrinsic value – that is, nature has value in and of itself, "a value neither conferred nor revocable" (Soulé 1985: 731) – and that this value system alone should drive conservation efforts. And, what is the primary tool to operationalize this view? The establishment of areas devoted to biodiversity, and the concomitant allocation of resources to protect these spaces. The National Park concept, for which Yellowstone National Park serves as the model, is, historically, a Western-based idea with the goal of strict nature protection (Stevens 1997). Following this model, in 1978, the IUCN first established a set of categories, objectives, and criteria for protected areas that could be implemented worldwide. The current system, which came into place in 1994, differs from this initial system in a number of important ways: it provides a definition for what constitutes a protected area;[10] the number of protected area categories was reduced from 10 to six; and, it allows for a "gradation of human intervention," whereby the protected area categories vary in the degree to which they recognize resident peoples' settlement and subsistence practices and their involvement in natural resource management.[11] While these categories clarify that not all protected areas are the same, the management regime of those categories that do emphasize strict protection have been interpreted as: if you are a human, you are either a protector of nature or you are threat to it (see Kareiva & Marvier 2012).

Since conservation biologist Michael Soulé's (1985) formative paper that defined the field of Conservation Biology, there has been ongoing discussion and building debate among conservation biologists about what counts as "nature" and how to best protect it. In recent years, scholarly exchange on these issues has re-surfaced, this time concerning the future of biodiversity conservation. Conservation biologists Peter Kareiva and Michelle Marvier (2012) articulated what they see as an alternative view of conservation; one that aims to be more attuned to the current global context in which we currently live and practice conservation (i.e., significantly larger global human population and increased rates of consumption).[12] Their primary concern with the traditional view of conservation is its "inattention to human well-being" (p. 963).[13] They contend that because actions taken to preserve biodiversity not only affect biodiversity but also people's lives and livelihoods, we need to take a more integrated approach – one that "jointly maximizes benefits to people and to biodiversity" (p. 962). Further, they challenge the idea of "pristine nature," encourage attention to multiple value systems and working with a diverse set of partners (i.e., including corporations), and advocate for greater attention to conservation in anthropogenic landscapes. This is not to say that maintaining protected areas is no longer a goal, for there is substantial evidence that this type of conservation action can be effective. For example, animal ecologist Francesco Rovero and colleagues (2015) found that populations of three species of primates in the Udzungwa Mountains, Tanzania remained stable in protected areas but declined in unprotected ones. Instead, the argument is two-fold: 1) we need to be cognizant that protected areas often negatively impact the communities living around them, and that 2) focusing our efforts in *only* these spaces may no longer be enough to prevent biodiversity. With data from 60 protected areas from across the world, biologist William Laurance and colleagues (2014), for example, determined that while half of the reserves were deemed effective (or passably so), the remaining suffered from activities occurring in the matrix surrounding them, thereby suggesting the importance of extending the purview of focus beyond the reserves themselves.

While it is heartening to see these dialogues and debates emerge from the field of conservation biology itself, let us not forget that anthropology and other social science and humanistic fields have also been at the forefront of addressing the questions of conservation "of what" and "for whom." Having a home in (or even an alliance with) anthropology means that primatologists can not only draw conceptually from the anthropological discourse on the concept of nature and the practice of conservation, but also draw methodologically from the rich set of tools anthropology offers to engage, understand, and convey human experience. Indeed, "conservation requires people skills" (Caro &Sherman 2013: 309); primatology's anthropological base provides us access to those skills.[14]

Lessons from the Environmental Social Sciences and Humanities

Anthropologists and other social science and humanities scholars have established that humans have long played an active role in altering and maintaining what it is that we call "nature."[15] It is important that we not misunderstand the motivation of this critique, nor dismiss it as mere "ivory tower" business; the implications of it are profound and real. By problematizing the idea of "nature," the goal is not to diminish the importance of biodiversity conservation, but rather, to demonstrate that the way nature is perceived (e.g., nature is untouched by humans; nature is constant and stable) determines the way it is managed (Botkin 1990).[16] Namely, viewing nature as separate from humans can often translate into conservation strategies that emphasize human exclusion. For example, many protected areas have been created out of lands with long histories of human occupancy and use (Neumann 1998). Exclusion can involve the eviction of people from their land, but also when people are restricted in their use of the area and are excluded from decision-making regarding its management (e.g., local Maasai and the Manyara Ranch Conservation Area in Tanzania; Goldman 2011).

A related critique, also with important implications for the practice of conservation, is that how we answer the question "why preserve biodiversity?" depends on our value system. Environmental studies scholar Richard Peterson and colleagues (2010) argue that conservation practitioners tend to approach the question in ways that are "rooted in the values, perceptions, and methods of Western conservation science and culture" (p. 7). That is, preserving biodiversity means, for example, protecting wild places, preventing habitat degradation, and thwarting species extinction. People who inhabit biodiverse rich areas and live adjacent to protected areas, however, may not share the same perceptions of natural resources and conservation values (as Western conservationists). For example, anthropologists Conrad Kottak and Alberto Costa (1993) have argued that in Madagascar the economic value of forest (e.g., watershed protection) is a much more powerful incentive against forest degradation than are global goals like "preserving biodiversity" or lemur conservation. There also may be diversity *within* a community in terms of beliefs, values, and attitudes toward nature and its conservation.

Accordingly, another important contribution of anthropologists, and other social scientists, has been to critically examine the implications of concepts such as, "local," "community," and "indigenous," – concepts that inundate the discourse on protected area conservation.[17] The notion of "community" often ignores the diverse interests and processes within communities, conveying a false sense of homogeneity (Agrawal & Gibson 1999). These concerns are particularly salient in areas that have witnessed considerable migration where a resulting diversity in backgrounds (i.e., ethnicities, lifestyles, histories, and ecologies) may translate into differing views, concerns, and interests with regard to forest use and protection (Ostrom 1999; Davis 2011). For example, in the highlands of Central Sulawesi, I explored local perceptions of nature and attitudes toward protected area conservation by conducting semistructured interviews with community members of the village of Tomado.[18] I found differences between the indigenous *To Lindu* and migrants in the way the forest is conceptualized: while the Lindu respondents see the forest as a source of livelihood (e.g., providing forest products, such as rattan and firewood) and speak to the value of its persistence for future generations (e.g., providing watershed protection), many migrants spoke of the forest in terms of what it could be turned into (i.e., additional agricultural land) rather than as a space to be preserved (Riley 2013).

So how might the conservation social science being done by our fellow anthropologists be of mutual interest to and benefit primatologists involved in advancing conservation?[19] The first lesson is that how we think about primates, nature, and humans' place in it shapes how we approach conservation interventions. By holding on to a pristine "primates in nature" perspective, we become unable to see beyond the "conservation in totally protected areas" mindset and approach (cf. Struhsaker 1998). This is not to say that we should forego efforts to maintain intact, minimally altered habitats for threatened primates; but rather, that protected areas alone are not enough. Only a small amount of forested area actually falls within protected areas (i.e., 12 percent of Earth's land surface: Chape et al. 2005) and many remaining populations of threatened primates do not necessarily occur within them. Seventy-five percent of all orangutans, for example, live outside of protected areas, so our ability to develop effective conservation plans is hampered if we don't have a good understanding of whether and how they are able to persist in human-modified areas (Meijaard et al. 2010; Spehar et al. 2018). Moreover, in the current era in which human-nonhuman primate sympatry is becoming increasingly widespread, traditional forest and wildlife protection strategies that emphasize human exclusion or ignore anthropogenic environments may no longer be effective in protecting primate populations (Chapman & Peres 2001), particularly when sympatry results in competition and conflict.

Primate conservation in the twenty-first century and beyond therefore means broadening the realm of our foci and toolkits (Figure 7.2), and excitingly, many scholars have been and are paving the way.[20] For example, conservation efforts should focus on evaluating the conservation value of anthropogenic environments,

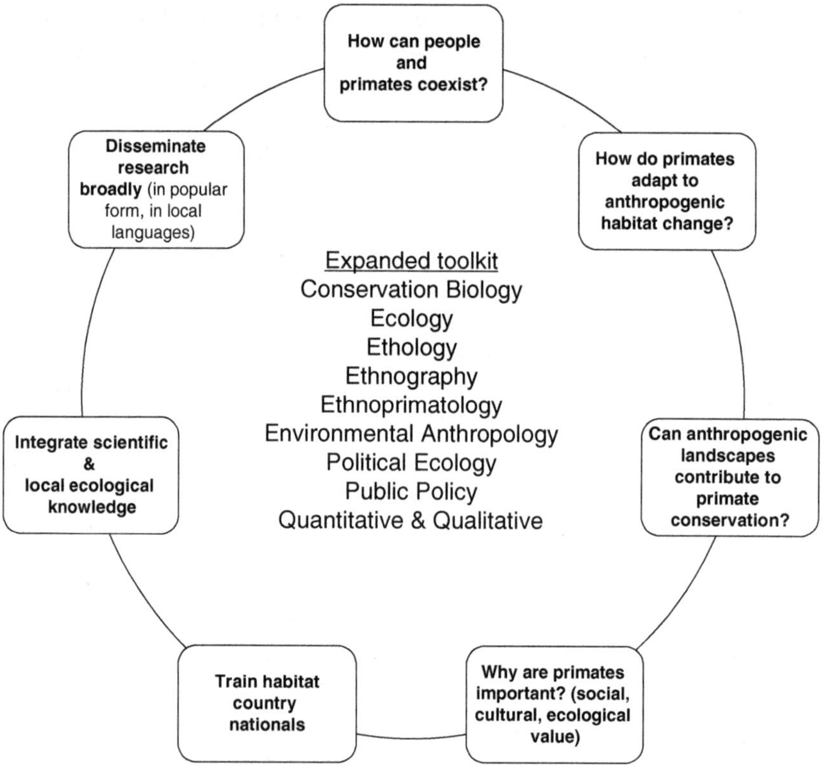

FIGURE 7.2 An expanded toolkit for primate conservation. The traditional approach to primate conservation emphasizes the expansion and protection of reserves, monitoring, and education. Primate conservation in the twenty-first century and beyond necessitates an expanded lens and toolkit. Such an approach maintains a fundamental concern with the protection and persistence of primate populations, but does so in a way that is attuned to contemporary realities (e.g., primates increasingly exist in anthropogenic spaces) and issues of social justice.

such as agroforesty systems (Estrada et al. 2012), expanding research beyond protected areas (Wich & Marshall 2016), developing the conservation capacity of habitat country nationals (Trevelyan & Nuttman 2008), and evaluating how humans and primates can sustainably coexist (Humle & Hill 2016; McLennan et al. 2017; Nekaris et al. 2017). "Broadening the realm" also means recognizing that conservation is not merely a biological problem concerned with forests, trees, or primates. Conservation is fundamentally a human endeavor; it is conceived, designed, and enacted by people. It is therefore also a social problem (Estrada 2013). Accepting this fundamental reality enables us to appreciate how, for example, many cases of human-wildlife conflict also constitute conflicts between different groups of people *about* wildlife. For example, as I discussed in Chapter 6, there is an introduced, free-ranging population of rhesus macaques (*Macaca mulatta*) that lives along the Silver River in Ocala, Florida and park officials are increasingly concerned about the potential for aggressive physical contact and disease transmission (Riley & Wade 2016). This interface may be characterized as an example of human-wildlife conflict; however, this label can mask "the complex and dynamic nature of the underlying 'conflict' that stems from differential values, needs, priorities, and power relations between the human groups concerned" (Hill 2015: 3).[21] In other words, it makes the "human-human conflict" at play invisible, for there also is conflict between different groups of people (e.g., state and park officials versus local residents) about the presence of the macaques and how to manage them.[22]

Conservation is a social, cultural, and political process that is inextricably linked with issues of social justice and equity (Alcorn 2005; Brechin et al. 2002; Brosius 2006). For primatologists, this means that we need to recognize the moral complexity of our fieldwork; our ethics encompasses the intersection of human, nonhuman primate, and ecosystem needs and interests (Riley & Bezanson 2018). In doing so, we must consider how to not only conserve our primate subjects and their habitats, but also how our research results and conservation efforts may affect, negatively or positively, the human communities that surround our field sites (Hill 2002; MacKinnon & Riley 2010), particularly for sites such as protected area that have "deep social effects" on people living in and around them (West & Brockington 2006:611). To do so, we need to adopt "reflexive empiricism," if you will, and, as addressed above, we need people skills. Sociocultural anthropologists have the knowledge and skills needed to understand how and why different groups of people conceptualize and interact with the environment in the ways that they do. Having a home in anthropology means that we can forge collaborations with anthropologists who embody these skills. For example, biological anthropologist, Melissa Remis, and sociocultural anthropologist, Rebecca Hardin, integrate their respective expertise to better understand human-wildlife interactions in the Dzanga-Sangha Dense Forest Reserve (RDS) in the Central African Republic and the consequences of these interactions for conservation. Their collaborative, intra- and interdisciplinary, mixed-methods approach enabled them to determine whether and how gorillas use anthropogenically altered areas, and explain how

FIGURE 7.3 Adult female Western lowland gorilla with juvenile (*Gorilla gorilla gorilla*) in the Dzanga Sangha Protected Areas Complex (APDS), Central African Republic. Photo credit: Tiffany Gobert.

people's response to the gorillas is linked to the varying ways gorillas are valued – i.e., what they call "transvaluation": gorillas have value for tourism, for bushmeat, and as conservation icons (Figure 7.3). This type of collaborative work can then help foster more socially inclusive approaches to the monitoring and conservation management of gorillas in the RDS (Hardin & Remis 2006; Remis & Hardin 2009).

In addition to collaborating with other anthropologists, our anthropological home base also provides opportunities to develop these skills ourselves. This is a very special position to be in that we should not take lightly. The integration of conservation biology and the social sciences, more generally, has experienced difficulties largely stemming from communication barriers (i.e., deciphering disciplinary jargon), epistemological and methodological differences (i.e., reconciling positivist and interpretivist approaches), lack of mutual respect, limited opportunities for collaboration due to discipline-focused structuring of funding and institutions, and differences in publishing norms (Campbell 2005; Drew & Henne 2006; Fox et al. 2006). Anthropological primatologists are in a unique position in that they generally work within the nomothetic tradition of the natural sciences, yet they have been imbued with the interpretive, meaning-seeking approaches of anthropology through academic coursework they may have taken (even if reluctantly) and by being faculty members in anthropology departments. Some boundary crossing must, nonetheless, still take place, and the largest hurdle is likely methodological. Primatologists generally conduct hypothesis-driven research, collecting quantitative data that are amenable to statistical analysis in order to draw generalizable conclusions. Because of this tradition, when a human component is incorporated into a project, quantitative techniques such as questionnaires and surveys generally predominate. This is not to say that these techniques have no value; rather, the concern is that a questionnaire does little to expose the deeper meaning behind people's beliefs, attitudes, and actions toward primates and conservation. In other words, primatologists need to appreciate the value of qualitative methods – e.g., participant observation, in-depth open-ended interviews – and the data that ensue from these methods.[23]

The good news is that, as we saw from Chapter 6, this is happening. And, more recently, anthropologist Kerry Dore and colleagues (2017) published an edited volume that provides practical guidance on how to integrate traditional primatological methods with qualitative research methods, such as in-depth interviews, participant observation, cultural mapping, and human focal follows. Joanna Setchell and colleagues (2017) also argue for integrating biological and social methods – what they call a "biosocial approach" – to address issues related to human-primate coexistence and highlight case studies exemplifying this approach.[24] For example, anthropologist Siân Waters involved local shepherds from the Bouhachem Forest, Northern Morocco in the research and integrated their local ecological knowledge (accessed through ethnographic fieldwork) with primatological data to co-produce knowledge on the population status and ecology of Barbary macaques (*Macaca sylvanus*) at the site (Figure 7.4). In doing so, shepherds became invested

FIGURE 7.4 A shepherd and his goat herd in Bouhachem forest, North Morocco. Photo credit: Siân Waters.

in the project and Waters witnessed a shift in their behavior toward the macaques from hunting to protection.

To advance this trend of integrative biological and social research requires a commitment by primatologists. If a critical component of ethically-engaged primatology is to pay attention to the conservation of the primates and habitats where we work (Nekaris & Nijman 2013; Riley & Bezanson 2018), then it is also our ethical responsibility to ensure that the next generation of primatologists is equipped with the knowledge and skills needed to understand and address the human dimension of primate conservation. It is therefore the responsibility of senior primatologists to encourage and facilitate intradisciplinary training (i.e., do not advise students to avoid sociocultural anthropology courses), and the responsibility of junior and budding primatologists to seek it out (i.e., do not avoid taking said courses).[25]

Notes

1 The Primate Specialist Group (PSG) was organized by Russell Mittermeier in 1977. It is one of more than 140 Specialist Groups that comprise the IUCN Species Survival Commission (IUCN/SSC). The IUCN/SSC is a global alliance of conservationists that is tasked with providing the IUCN information on "biodiversity conservation, the inherent value of species, their role in ecosystem health and functioning, the provision of ecosystem services, and their support to human livelihoods" (www.iucn.org/theme/species/about/species-survival-commission). This information is, in turn, used to develop the IUCN Red List of Threatened Species (www.iucnredlist.org/).

2 Human population data: United Nations, Department of Economic and Social Affairs, Population Division (2017). World Population Prospects: The 2017 Revision, custom data acquired via website. The Food and Agriculture Organization of the United Nations (FAO) estimates that between 1980–1990 there was a loss of 9.9 million hectares of forest and other wooded lands per year, and a net loss of 129 million ha of forest between 1990–2015 (www.fao.org/forest-resources-assessment/past-assessments/en/).

3 FAO (2016) Food and Agriculture Organization of the United Nations. The State of the World's Forest 2016. UNDES (2017) United Nations Department of Economic and Social Affairs. World Population Prospects: The 2017 Revision. United Nations, New York.

4 To combat this "strong selective pressure," Chapman and Peres (2001: 24) called for the editors of high-impact journals to give due attention to high-quality papers with a conservation focus.

5 The list could surely go on and on. Here I draw from Chapman and Peres (2001), Angeloni et al. (2008), Caro and Sherman (2013); Oates (2013).

6 Primatologist John Oates (2013) conducted a targeted survey where he asked 14 "wise elders," as he calls them, for their views on how to better effect conservation. Based on their own experiences in conservation, these respondents identified what they saw as the major threats to primate populations and offered their perspectives on how to combat them. While the expertise of his set of respondents is valued, a follow up study could be to systematically survey a more diverse set of primatologists – i.e., diverse in terms of the demographics of the respondents (i.e., age, sex, nationality, and ethnic background) and their perspectives toward conservation.

7 See Caro (2007), Agenloni et al. (2008), Caro and Sherman (2013), Berger-Tal et al. (2016). For examples of how scientific research in general has contributed to conservation action, see Wrangham and Ross (2008).

8 It seems though that Caro's perspective has changed since this 2007 paper. For instance, in challenging reason number 9 animal behaviorists avoid conservation ("I am an evolutionary behaviorist, and I don't see the relevance of evolution to conservation," Tim Caro and Paul Sherman (2013: 308) counter with support that "anthropogenic change provides new opportunities to derive and test theoretical predictions about adaptations and constraints." They also challenge the claim that conservation biology is not rich in theory by pointing out important contributions from island biogeography and economic theory.

9 See Tilmam (2012) for a review.

10 A protected area is "an area of land and/or sea especially dedicated to the protection and maintenance of biological diversity, and of natural and associated cultural resources, and managed through legal or other effective means"; Phillips 2004: 10).

11 The six categories are: Ia. Strict Nature Reserve, Ib. Wilderness Area, II. National Park, III. National Monument or Feature, IV. Habitat/Species Management Area, V. Protected Landscape/Seascape, VI. Protected Area with sustainable use of natural resources (www.iucn.org/theme/protected-areas/about/protected-areas-categories). It is worth noting that the rhetoric accompanying this system reveals that human inclusion is recognized only to the extent that it does not conflict with the primary goal of conservation of biodiversity (e.g., an objective of Category II. National Park protected area is "to take into account the needs of indigenous people, including subsistence resource use, in so far as these will not adversely affect the other objectives of management," Dudley 2008: 16; see also Stevens 1997).

12 Their view has been dubbed "new conservation" – a name that in itself has been criticized (i.e., "what they advocate is not conservation"; Soulé 2013; "there's nothing new to their perspective"; Greenwald et al. 2013; Soulé 2013). That said, the authors (Peter Kareiva and Michelle Marvier) actually never called it this.

13 Critics (e.g., Greenwald et al. 2013) argue that concern for human well-being has always been a part of the traditional view of conservation, citing Soulé's claim that "any recommendations about the location and size of national parks should consider the impact of the park on indigenous peoples and their cultures, on the local economy, and on opportunity costs such as forfeited logging profits" (Soulé 1985: 727–28). I have two things to say about this: 1) this claim limits concern for human well-being to traditional peoples to the exclusion of millions of other people who may not count as "indigenous" and yet are deeply affected by the creation of protected areas (e.g., displacement, loss of source of livelihood, etc.); 2) how can an approach that privileges just one value system – the intrinsic value of biodiversity – address these impacts, for biodiversity will always win.

14 Phyllis Dolhinow (2002: 21), a pioneer of field primatology, recognized this: "Whatever our study location, New or Old World, an anthropological perspective will be a powerful force affecting the practice of primatology by providing additional theoretical foci and sound advice concerning ways to effect conservation."

15 See, for example, Adams and McShane (1992), Cronon (1995), Kealhofer (2003), Morin-Rivat et al. (2014).

16 Michael Soulé (1995) wrote forcefully about what he called "the social siege of nature" – one prong of the attack on nature led by social scientists who, as part of the postmodern turn, challenged the idea of nature, arguing that nature, and related

concepts such as wilderness and biodiversity, are mere human constructions, and that the preservation of so-called nature is an activity of the privileged done at the expense of the marginalized. While indeed denying the objective reality of nature has been part of the constructivist perspective (Escobar 1996), which natural scientists undoubtedly see little value in, the critique has also generated important insight which can actually improve (and has improved) the process of biodiversity conservation. I review some of this insight herein.

17 See for example: Zerner (1994), Li (1996), Forsyth (1996), Brosius et al. (1998), Brosius (1999), King (2007), Flint et al. (2008).

18 The village of Tomado is inhabited by indigenous *To Lindu*, other Kaili people from nearby lowland villages, and Bugis people from South Sulawesi, Indonesia.

19 It is worth noting that the flow of impact is not unidirectional. Shoreman-Ouimet and Kopnina (2016) call for an anthropology of conservation "that acknowledges the agency of nonhuman subjects in political ecologies of cohabitation and integrates interspecies ethics, ethology, and conservation ecology into its methodology" (p. 16). As we have seen in previous chapters, ethnoprimatology has done this, and could serve as a model.

20 Here I specifically acknowledge the work of Shirley Strum (1984) and David Western (1984), who have been advocating for a more inclusive (of humans) approach to primate conservation for quite some time. For example, Strum (184: 380) wrote: "What should be apparent by now to those interested in primate conservation in source countries is that the 'people factor' cannot be ignored or dismissed." In addition, while Russell Mittermeier's first edition of the *Global Strategy for Primate Conservation* (1978) primarily emphasizes the creation and protection of reserves free from human influence, it does call for identifying ways for humans and primates to coexist in multiple-use areas.

21 During geographer Mara Goldman's (2011) research on the politics of conservation in Tanzania, a Maasai elder told her: "I do not hate wildlife and not want them here. Nor do I like them for tourists because we don't see them [tourists]. Wildlife on the side of tourists is not the problem. It is the people who protect the wildlife that are the problem" (p. 76). Furthermore, in May 2009, on the western side of Mount Kilimanjaro in Tanzania, a large group of villagers chased a herd of elephants toward a cliff; six of the elephants died. Based on ethnographic research with villagers, environmental studies scholar Sayuni Mariki and colleagues (2015) attribute the primary cause of this act of violence to frustration aimed not at the elephants per se, but at governmental authorities for their lack of effort in addressing human-elephant conflict (e.g., positioning park officials to deter crop feeding elephants) and for imposing conservation practices that marginalize the villagers. Further examples abound: see Peterson et al. (2010); Redpath et al. (2013); Madden and McQuinn (2014).

22 The Silver River macaques are deemed invasive, exotic animals unworthy of protection by wildlife and park officials, yet they have achieved a special place in the hearts of boaters who frequent the Silver River (Riley 2019).

23 Here I draw from broader conversations on the value of qualitative social science for conservation. See Peterson et al. (2010), Davis (2011), Rust et al. (2017). Anthropologist John Knight's (2017) introduction to the section on incorporating ethnography in primatological research in Dore et al. (2017) *Ethnoprimatology: A Practice Guide to the Research at the Human-Primate Interface* also provides valuable insight on how ethnographic fieldwork for a primatologist is "not simply a new *method* of inquiry, but also

a new *focus* of inquiry" (p. 171), and how ethnographic fieldwork is not just about conducting interviews; it requires deep immersion in the lives of the communities with whom we are working.

24 This paper is particularly heartening given that the authors reveal that four out of five of them attribute their academic backgrounds to the biological sciences.

25 This may be more difficult in countries where primatology is not linked to anthropology (e.g., Vietnam; Hoàng 2016). The international primatology community can help remedy this by providing training opportunities, such as the pre-congress workshops that accompany the bi-annual International Primatological Society meetings.

References

Adams J, McShane T. 1992. *The Myth of Wild Africa*. Berkeley: University of California Press.

Agrawal A, Gibson CC. 1999. Enchantment and disenchantment: The role of community in natural resource management. *World Development* 27: 629–49.

Alcorn J. 2005. Dances around the fire: Conservation organizations and community-based natural resource management. In *Representing Communities: Histories and Politics of Community-Based Natural Resource Management*, ed. JP Brosius, A Tsing, C Zerner.

Angeloni L, Schlaepfer MA, Lawler JJ, Crooks KR. 2008. A reassessment of the interface between conservation and behaviour. *Animal Behaviour* 75: 731–37.

Berger-Tal O, Blumstein DT, Carroll S, Fisher RN, Mesnick SL, et al. 2016. A systematic survey of the integration of animal behavior into conservation. *Conservation Biology* 30: 744–53.

Botkin D. 1990. *Discordant Harmonies: A New Ecology for the Twenty-First Century*. Oxford: Oxford University Press.

Brechin SR, Wilshusen PR, Fortwangler CL, West PC. 2002. Beyond the square wheel: Toward a more comprehensive understanding of biodiversity conservation as social and political process. *Society &Natural Resources* 15: 41–64.

Brosius JP. 1999. Green dots, pink hearts: displacing politics from the Malaysian rain forest. *American Anthropologist* 101: 36–57.

Brosius JP. 2006. Common ground between anthropology and conservation biology. *Conservation Biology* 20: 683–85.

Brosius JP, Tsing A, Zerner C. 1998. Representing communities: Histories and politics of community-based natural resources management. *Society and Natural Resources* 11: 157–68.

Bublitz DC, Wright PC, Rasambainarivo FT, Arrigo-Nelson SJ, Bodager JR, Gillespie TR. 2015. Pathogenic enterobacteria in lemurs associated with anthropogenic disturbance. *American Journal of Primatology* 77: 330–37.

Buchholz R. 2007. Behavioural biology: An effective and relevant conservation tool. *Trends in Ecology & Evolution* 22: 401–07.

Calvignac-Spencer S, Leendertz SAJ, Gillespie TR, Leendertz FH. 2012. Wild great apes as sentinels and sources of infectious disease. *Clinical Microbiology and Infection* 18: 521–27.

Campbell LM. 2005. Overcoming obstacles to interdisciplinary research. *Conservation Biology* 19: 574–77.

Caro T. 2007. Behavior and conservation: A bridge too far? *Trends in Ecology & Evolution* 22: 394–400.

Caro T, Sherman PW. 2013. Eighteen reasons animal behaviourists avoid involvement in conservation. *Animal Behaviour* 85: 305–12.

Chape S, Harrison J, Spalding M, Lysenko I. 2005. Measuring the extent and effectiveness of protected areas as an indicator for meeting global biodiversity targets. *Philosophical Transactions of the Royal Society B: Biological Sciences* 360: 443–55.

Chapman CA, Peres CA. 2001. Primate conservation in the new millennium: The role of scientists. *Evolutionary Anthropology* 10: 16–33.

Chiyo PI, Moss CJ, Alberts SC. 2012. The influence of life history milestones and association networks on crop-raiding behavior male African elephants. *PLOS One* 7: 1–11.

Cozzi G, Chynoweth M, Kusak J, Çoban E, Çoban A, et al. 2016. Anthropogenic food resources foster the coexistence of distinct life history strategies: Year-round sedentary and migratory brown bears. *Journal of Zoology* 300: 142–50.

Cronon W. 1995. *Uncommon Ground: Toward Reinventing Nature*. New York: WW Norton & Company.

Davis A. 2011. 'Ha! What is the benefit of living next to the park?' Factors limiting in-migration next to Tarangire National Park, Tanzania. *Conservation and Society* 9: 25–34.

Dore KM, Riley EP, Fuentes A. 2017. *Ethnoprimatology: A Practical Guide to Research at the Human–Nonhuman Primate Interface*. Cambridge: Cambridge University Press.

Drew JA, Henne AP. 2006. Conservation biology and traditional ecological knowledge: integrating academic disciplines for better conservation practice. *Ecology and Society* 11: 34.

Dudley N, ed. 2008. *Guidelines for Applying Protected Area Management Categories*. Gland, Switzerland: IUCN.

Escobar A. 1996. Constructing nature: Elements for a poststructural political ecology. In *Liberation Ecologies: Environment, Development and Social Movements*, ed. R Peet, M Watts, pp. 46–67. New York: Routledge.

Estrada A. 2013. Socioeconomic contexts of primate conservation: Population, poverty, global economic demands, and sustainable land use. *American Journal of Primatology* 75: 30–45.

Estrada A, Garber PA, Rylands AB, Roos C, Fernandez-Duque E, et al. 2017. Impending extinction crisis of the world's primates: Why primates matter. *Science Advances* 3: e1600946.

Estrada A, Raboy BE, Oliveira LC. 2012. Agroecosystems and primate conservation in the tropics: A review. *Amercan Journal of Primatology* 74: 696–711.

Fairgrieve C. 1995. *The comparative ecology of blue monkeys (Cercopithecus mitis stuhlmannii) in logged and unlogged forest, Budongo Forest Reserve, Uganda: The effects of logging on habitat and population density*. University of Edinburgh.

FAO. 2016. Food and Agriculture Organization of the United Nations. The State of the World's Forest 2016.

Flint CG, Luloff AE, Finley JC. 2008. Where is "community" in community-based forestry? *Society & Natural Resources* 21: 526–37.

Forsyth T. 1996. Science, myth and knowledge: testing the theory of Himalayan environmental degradation in Thailand. *Geoforum* 27: 375–92.

Fox HE, Christian C, Nordby JC, Pergams OR, Peterson GD, Pyke CR. 2006. Perceived barriers to integrating social science and conservation. *Conservation Biology* 20: 1817–20.

Goldman MJ. 2011. Strangers in their own land: Maasai and wildlife conservation in Northern Tanzania. *Conservation and Society* 9: 65–79.

Green SJ, Armstrong J, Bogan M, Darling E, Kross S, et al. 2015. Conservation needs diverse values, approaches, and practitioners. *Conservation Letters* 8: 385–87.

Greenwald N, Dellasala DA, Terborgh JW. 2013. Nothing new in Kareiva and Marvier. *BioScience* 63: 241–41.

Hardin R, Remis MJ. 2006. Biological and cultural anthropology of a changing tropical forest: A fruitful collaboration across subfields. *American Anthropologist* 108: 273–85.

Hill CM. 2002. Primate conservation and local communities--ethical issues and debates. *American Anthropologist* 104: 1184–94.

Hill CM. 2015. Perspectives of "Conflict" at the wildlife–agriculture boundary: 10 years on. *Human Dimensions of Wildlife* 20: 296–301.

Hoàng TM. 2016. Development of primatology and primate conservation in Vietnam: Challenges and prospects. *American Anthropologist* 118: 130–37.

Humle T, Hill CM. 2016. People-primate interactions: Implications for primate conservation. In *Introduction to Primate Conservation*, ed. AJ Marshall, SA Wich, pp. 219–40. Oxford: Oxford University Press.

Janson CH. 2000. Primate socio-ecology: The end of a golden age. *Evolutionary Anthropology* 9: 73–86.

Kareiva P, Marvier M. 2012. What is conservation science? *BioScience* 62: 962–69.

Kealhofer L. 2003. Looking into the gap: Land use and the tropical forests of southern Thailand. *Asian Perspectives* 42: 72–95.

King BH. 2007. Conservation and community in the new South Africa: A case study of the Mahushe Shongwe Game Reserve. *Geoforum* 38: 207–19.

Knight J. 2017. Introduction to Part II. In *Ethnoprimatology: A Practical Guide to Research at the Human-Primate Interface*, ed. KM Dore, EP Riley, A Fuentes, pp. 171–75. Cambridge: Cambridge University Press.

Kottak CP, Costa ACG. 1993. Ecological awareness, environmentalist action, and international conservation strategy. *Human Organization* 52: 335–43.

Krief S, Cibot M, Bortolamiol S, Seguya A, Krief J-M, Masi S. 2014. Wild chimpanzees on the edge: Nocturnal activities in croplands. *PLoS ONE* 9: e109925.

Laurance WF, Useche DC, Rendeiro J, Kalka M, Bradshaw CJ, et al. 2012. Averting biodiversity collapse in tropical forest protected areas. *Nature* 489: 290.

Li TM. 1996. Images of community: Discourse and strategy in property relations. *Development and Change* 27: 501–27.

MacKinnon K, Riley EP. 2010. Field primatology of today: Current ethical issues. *American Journal of Primatology* 72: 749–53.

Madden F, McQuinn B. 2014. Conservation's blind spot: The case for conflict transformation in wildlife conservation. *Biological Conservation* 178: 97–106.

Mariki SB, Svarstad H, Benjaminsen TA. 2015. Elephants over the cliff: Explaining wildlife killings in Tanzania. *Land Use Policy* 44: 19–30.

Meijaard E, Albar G, Rayadin Y, Ancrenaz M, Spehar S. 2010. Unexpected ecological resilience in Bornean orangutans and implications for pulp and paper plantation management. *PLoS One* 5: e12813.

McLennan MR, Spagnoletti N, Hockings KJ. 2017. The implications of primate behavioral flexibility for sustainable human–primate coexistence in anthropogenic habitats. *International Journal of Primatology* 38: 105–21.

Mittermeier RA. 1978. A global strategy for primate conservation. *New York: IUCN-SSC Primate Specialist Group and the New York Zoological Society.*

Morin-Rivat J, Fayolle A, Gillet J-F, Bourland N, Gourlet-Fleury S, et al. 2014. New evidence of human activities during the holocene in the lowland forests of the Northern Congo Basin. *Radiocarbon* 56: 209–20.

Nekaris K, Poindexter S, Reinhardt K, Sigaud M, Cabana F, et al. 2017. Coexistence between Javan slow lorises (Nycticebus javanicus) and humans in a dynamic agroforestry landscape in West Java, Indonesia. *International Journal of Primatology* 38: 303–20.

Nekaris KA-I, Nijman V. 2013. Complex and heterogeneous ethical structures in field primatology. In *Ethics in the Field: Contemporary Challenges*, pp. 108–23. New York: Berghahn.

Neumann RP. 1998. *Imposing Wilderness: Struggles over Livelihood and Nature Preservation in Africa*. Berkeley: University of California Press.

Oates JF. 2013. Primate conservation: Unmet challenges and the role of the International Primatological Society. *International Journal of Primatology* 34: 235–45.

Ostrom E. 1999. Self-governance and forest resources. *CIFOR Occasional Papers*: 1–14.

Peterson MN, Birckhead JL, Leong K, Peterson MJ, Peterson TR. 2010. Rearticulating the myth of human–wildlife conflict. *Conservation Letters* 3: 74–82.

Peterson RB, Russell D, West P, Brosius JP. 2010. Seeing (and doing) conservation through cultural lenses. *Environmental Management* 45: 5–18.

Phillips A. 2004. The history of the international system of protected area management categories. *Parks* 14: 4–14.

Redpath SM, Young J, Evely A, Adams WM, Sutherland WJ, et al. 2013. Understanding and managing conservation conflicts. *Trends in Ecology & Evolution* 28: 100–09.

Remis MJ, Hardin R. 2009. Transvalued species in an African forest. *Conservation Biology* 23: 1588–96.

Riley EP. 2013. The human–macaque interface in the Sulawesi Highlands. In *The Politics of Species: Reshaping our Relationships with Other Animals*, ed. R Corbey, A Lanjouw, pp. 189–96: Cambridge University Press.

Riley EP. 2019. Towards tolerance and coexistence: A comparative analysis of the human-macaque interface in Sulawesi, Indonesia and Florida, USA. In *Human-Wildlife Interactions: Turning Conflict into Coexistence*, ed. B Frank, JA Glikman, S Marchini, pp. 198–215. Cambridge: Cambridge University Press.

Riley EP, Bezanson M. 2018. Ethics of primate fieldwork: Toward an ethically-engaged primatology. *Annual Review of Anthropology*.

Riley EP, Wade TW. 2016. Adapting to Florida's riverine woodlands: the population status and feeding ecology of the Silver River rhesus macaques and their interface with humans. *Primates* 57: 195–210.

Rovero F, Mtui A, Kitegile A, Jacob P, Araldi A, Tenan S. 2015. Primates decline rapidly in unprotected forests: Evidence from a monitoring program with data constraints. *PLOS ONE* 10: e0118330.

Rust NA, Abrams A, Challender DW, Chapron G, Ghoddousi A, et al. 2017. Quantity does not always mean quality: The importance of qualitative social science in conservation research. *Society & Natural Resources* 30: 1304–10.

Schweitzer C, Gaillard T, Guerbois C, Fritz H, Petit O. 2017. Participant profiling and pattern of crop-foraging in chacma baboons (*Papio hamadryas ursinus*) in Zimbabwe: Why does investigating age–sex classes matter? *International Journal of Primatology* 38: 207–23.

Setchell JM. 2013. The top 10 questions in primatology. *International Journal of Primatology* 34: 647–61.

Setchell JM, Fairet E, Shutt K, Waters S, Bell S. 2017. Biosocial conservation: Integrating biological and ethnographic methods to study human–primate interactions. *International Journal of Primatology* 38: 401–26.

Shoreman-Ouimet E, Kopnina H. 2016. *Culture and Conservation: Beyond Anthropocentrism*. New York: Routledge.

Soulé M. 2013. The "New Conservation". *Conservation Biology* 27: 895–97.

Soulé ME. 1985. What is conservation biology? *BioScience* 35: 727–34.

Soulé ME. 1995. The social siege of nature. In *Reinventing Nature: Responses to Postmodern Deconstruction*, ed. ME Soulé, G Lease, pp. 137–70. Washington, DC: Island Press.

Spehar SN, Sheil D, Harrison T, Louys J, Ancrenaz M, et al. 2018. Orangutans venture out of the rainforest and into the Anthropocene. *Science Advances* 4: e1701422.

Stevens S. 1997. The legacy of Yellowstone, new alliances for conservation, lessons and directions. In *Conservation Through Cultural Survival: Indigenous Peoples and Protected Areas*, ed. S Stevens. DC: Island Press.

Struhsaker TT. 1998. A biologist's perspective on the role of sustainable harvest in conservation. *Conservation Biology* 12: 930–32.

Strum SC. 1984. Activist conservation: The human factor in primate conservation in source countries. In *Primate Ecology and Conservation*, ed. JG Else, PC Lee, pp. 367–82. Cambridge: Cambridge University Press.

Tilman D. 2012. Biodiversity & environmental sustainability amid human domination of global ecosystems. *Daedalus* 141: 108–20.

Trevelyan R, Nuttman C. 2008. The importance of training national and international scientists for conservation research. In *Science and Conservation in African Forests: The Benefits of Long-Term Research*, ed. R Wrangham, E Ross, pp. 88–98. Cambridge: Cambridge University Press.

UNDES. 2017. United Nations Department of Economic and Social Affairs. World Population Prospects: The 2017 Revision, United Nations, New York.

West P, Brockington D. 2006. An anthropological perspective on some unexpected consequences of protected areas. *Conservation Biology* 20: 609–16.

Western D. 1984. Introduction: Primate conservation in the broader realm. In *Primate Ecology and Conservation*, ed. JG Else, PC Lee, pp. 343–53. Cambridge: Cambridge University Press.

Wich SA, Marshall AJ. 2016. *An Introduction to Primate Conservation*. Oxford: Oxford University Press.

Willis KJ, Gillson L, Brncic TM. 2004. How "virgin" is virgin rainforest? *Science* 304: 402–03.

Wolf M, Van Doorn GS, Leimar O, Weissing FJ. 2007. Life-history trade-offs favour the evolution of animal personalities. *Nature* 447: 581.

Wrangham R, Ross E. 2008. *Science and Conservation in African Forests: The Benefits of Long-Term Research*. Cambridge: Cambridge University Press.

Zerner C. 1994. Through a green lens: the construction of customary environmental law and community in Indonesia's Maluku Islands. *Law & Society Review* 28: 1079–122.

8

CONCLUSION

Reclaiming Primatology as Anthropology

In the current era, most remaining populations of primates live in environments that have been influenced in some way by the anthropogenic touch: protected forests bisected by major roads, forest-farm edges, urban centers, and the riverine floodplains of a previously unknown land, to name a few. This book argues for a contemporary primatology that takes this reality seriously. Doing so means appreciating the value in studying primate adaptability at work – that is, uncovering how primates adapt to human-induced environmental change and theorizing outside-of-the-box to do so. In this book, I challenge the perspective that primates in anthropogenic contexts poses an epistemological and methodological dilemma for field primatology and showcase how an extended evolutionary approach enables us to rigorously examine primate behavior in these contexts. Namely, it allows us to explore *contemporary* evolution and expand our knowledge of the adaptive potential of our study species. It allows us to examine how primates and humans are co-shaping each other's ecology, sociality, and evolutionary trajectories. Ethnoprimatology, which I explore in Chapter 5, represents another outside-of-the-box approach. Its purview is not limited to just exploring primates in anthropogenic contexts, but also primates *with* humans in those spaces. It recognizes that our research objectives and practice cannot be disengaged from the human dimension. Accessing this dimension involves examining how humans and other primates behave together, the motivations and perceptions behind why humans view primates and interact with them in the ways that they do, and how our research and actions are aimed at conserving primates affects human lives and livelihoods. The multispecies commons that is the human-primate interface is complex and messy. Primatology needs anthropology to decode this glorious messiness. In this concluding chapter, I return to Washburn's promise of primatology to consider whether he would be pleased with what has become of primatology 45 years later, and how the promise of contemporary primatology that I articulate within compares. I also revisit one of the primary themes that underlie the promise of contemporary primatology: boundary crossing. Finally, I discuss the challenges that contemporary primatology faces moving forward.

Primatology's Promise: Then and Now

When Washburn published his *The Promise of Primatology* in 1973, he was excited to see a field that originally garnered only minor interest come of age, replete with its own scholarly associations and journals. As he saw it, primatology was not just about studying primates for the sake of studying primates, it was about studying primates to shed light on human nature. His promise for primatology was very much embedded in his vision for a dynamic, "New Physical Anthropology" he had articulated some 20 years prior. Washburn called for a problem-oriented primatology; that is, one guided by a robust theoretical framework: evolutionary theory. Washburn's envisioned a primatology that was experiment-oriented, as he considered experimentation to be the "only way to reliable knowledge and scientific progress" (p. 178). Washburn was also concerned about primatology developing in intellectual isolation and called for what he labelled an "eclectic primatology" – one that draws from multiple disciplines and employs the "methods of many different sciences" (p.182).

Forty-five years later, I would argue that we, too, have much to be excited about, but would Washburn be pleased with today's primatology? As we saw in Chapters 3 and 4, we have undoubtedly achieved his vision for a problem-oriented primatology. In line with Washburn's call for biological anthropology to engage more with evolutionary biology, primatology came to invoke evolutionary theory to explain the evolution of primate sociality and the variation we see in primate behavior via two major evolutionary forces, natural selection and sexual selection, in relation to the ecological environments in which primates live. But, as I have argued, primatology needs not stop there.[1] The contemporary contexts in which primates are living necessitate an extended evolutionary approach – one that recognizes the importance of behavioral plasticity as both enabling adaptability and driving evolutionary change. If behavioral plasticity enables primates to extend their range to include anthropogenic spaces, then it is possible that patterns of selection can also be altered. For example, the ability of primates to incorporate anthropogenic foods into their dietary repertoire may be adaptive in the short-term, but this plasticity may affect other fitness-related behavior, such as time away from building important social relationships with conspecifics, thereby altering the evolutionary trajectory of the species (Berger-Tal et al. 2011: 2). Further, if primates can incorporate human provisioned food into their diet and learn how to be with humans, are we (fieldworkers) effectively shifting selection regimes in favor of those individuals who are more tolerant of humans (Gumert et al. 2011)? An extended evolutionary approach for primatology also means recognizing and examining the central roles primates themselves play in the evolutionary process – e.g., as niche constructors, and in some cases, as co-niche constructors with humans.

As for Washburn's goal for an experiment-oriented primatology, I wouldn't exactly call contemporary behavioral primatology an experimental field. Indeed, field primatologists have developed innovative ways of: experimentally eliciting behaviors of interest (e.g., placing specific objects in the environment to observe stone handling; Leca et al. 2010) or to solve foraging tasks; Garber et al. 2012); understanding vocal repertoires through playback experiments (Cheney & Seyfarth 1982); and, testing spatial memory (Bicca Marques & Garber 2004). However, the difficulty in systematically controlling potentially confounding variables and the fact that we conduct our research in a more ethically-engaged manner today compared to Washburn's time means that the experimental extent of our practice will likely remain limited.[2] I would like to think that Washburn would be "OK" with that. That said, as we learned in Chapter 4, in the contemporary era "most wild populations are already serving as experimental subjects in the rapid changing environments caused by anthropogenic activities, which range from local and regional disturbance to global climate change" (Strier 2009: 225).

Washburn would most definitely be pleased with the interdisciplinary nature of today's primatology. Field primatologists employ theoretical and methodological approaches from ecology, ethology, biopsychology, evolutionary biology, and conservation biology. However, as I argue herein, the extent of our interdisciplinarity does not end there. In other words, the sources of interdisciplinarity need not and should not be confined to the natural sciences. This is where the promise I offer in this book likely deviates from Washburn's vision.[3] As we learned in Chapters 5 and 6, studying primates in anthropogenic settings necessitates attention to the human dimension, which in turn requires that we look to the social sciences and humanities for conceptual and methodological guidance. The ethnoprimatological approach, for example, is simultaneously integrative and interdisciplinary in nature, incorporating theory and method from primate ecology, disease ecology, sociocultural anthropology, environmental anthropology, conservation ecology, and more recently, human-animal studies, to study the human-primate interface. Ethnoprimatology's unit of investigation becomes the relationships that emerge as a result of the human-primate interface. Accordingly, it traverses beyond amorphous categories such as "anthropogenic impact" or "anthropogenic factors" to explore what the specific contexts are, how and why primates and people behave together, and the implications of these interconnections. This perspective is possible because of the anthropological base from which ethnoprimatology derives. This, in turn, is testament to the fact that contemporary primatology should never lose sight of its anthropological origin and what anthropology itself offers it.[4] Just as biological anthropologists George Armelagos and Dennis van Gerven (2003) argued for reclaiming biological anthropology as anthropology, the promise of contemporary primatology is about reclaiming primatology as anthropology (Table 8.1).

TABLE 8.1 Primatology as anthropology. American-British anthropologist MF Ashley-Montagu (1940) constructed a matrix to illustrate the connections between physical and social anthropology. Here I adapt and update his matrix to show how contemporary primatology aligns with sociocultural anthropology, thereby warranting reclaiming primatology as anthropology. Xs represent his original vision, and ✔s represent my updates. The question mark (?) represents where I deviate from Ashley-Montagu's perspective.

ANTHROPOLOGY	*Social Organization*	*Social Behavior*	*Individual Behavior*	*Linguistics*	*Material Culture*	*Archaeology*	*Religion*	*Mythology*	*Literature*	*Music*
Primatology	**X**	**X**	**X**	**X**	**X**	**X**	✔	✔	✔	**X -?**

Beyond Borders, Toward Pluralism

Washburn himself took issue with borders. He believed the departmental boundaries imposed by the structure of academic institutions stifled progress (Washburn 1968, 1973).[5] A central theme that permeates this book is the blurring and dissolving of boundaries – methodological, disciplinary, ontological, and epistemological. Primatology in the U.S. began in biological anthropology, initially borrowed methods from sociocultural anthropology, and then adopted theory and method from evolutionary biology and ethology. A key argument I make in this book is that there need not be a singular reason for primatology's being or only one way of doing primatology. Instead, the promise of contemporary primatology is about advancing distinct forms of hybridity. The increasingly predominant context in which today's primatology takes place is anthropogenic – human presence and activity play a role, either directly or indirectly, in shaping the environments in which primates live. These contexts are multispecies in distinct ways and unquestionably complex. This means a couple of important things. First, the boundaries that define our study subjects are shifting: primatology need not solely be the study of primates for implications for understanding of humans, but can also include humans as study subjects. This, in turn, forges an expanded purpose for primatology in anthropology. Second, standard evolutionary models may not be sufficient for capturing and explaining the complexity that is generated when humans and other primates interface.[6] In fact, by granting standard evolutionary theory a hegemonic status, we may be effectively closing ourselves off to other possibilities (Ingold 2007). For example, in discussing his fieldwork on the human-macaque interface in Bali, Indonesia, Agustín Fuentes (2010b: 610) writes:

> I suggest that the traditional language of evolutionary adaptation is, by itself, inadequate to explain the macaques' exploitation of human institutions and engineering projects. An ethnoprimatological approach places the ecology of the sites, the behavior of the humans and macaques, as well as historical and economic forces, into dynamic interaction. Actions and interactions by and between the macaques and humans are shaping and reshaping, constructing and coproducing, the shared components of ecological and social niches –and are doing so in ways that must be understood simultaneously through ethnographic, behavioral, and epidemiological lenses.

A useful parallel to help us think through this promise for contemporary primatology is the transformation human biology underwent when biological anthropologists realized that human biological adaptation cannot be fully understood without examining the social, cultural, economic, and political contexts in which it occurs.[7] Like human biology, contemporary primatology can be grounded in evolutionary theory, be open to extended evolutionary approaches, yet also be cognizant of context and contingencies. In other words, we need develop a culture of explanatory pluralism in primatology. What would this look like? First, it would mean *truly* recognizing the value in studying behavior at multiple levels of analysis. I say "truly" because descriptive research is typically looked down upon as less valuable, as less "scientific." And, yet, description and explanation are both important to the scientific process. As two pioneering field primatologists, Stuart Altmann and Jeanne Altmann, note: "The first without the second would be incomplete; the second could not exist without the first and is totally dependent on it for its veracity" (2003: 413). As we saw in Chapter 4, by studying primates in human-modified environments, we are asking important proximate questions about what the behavioral responses are in said environments, but we are also concerned with the adaptive significance – that is, the causes and consequences of behavioral flexibility. Doing both enables us to appreciate the complexity at play. For example, by "paying attention to the process, the natural history of events and individuals," Shirley Strum (2012: 7) was able to see "baboon complexity beyond social strategies." Strum observed her main study group of baboons in Kenya, "Pumphouse," fission. One hypothesis derived from socioecological theory is that it fissioned because it became too large. However, by focusing on "the natural history of events and individuals," she found that flexibility in foraging strategies in response to anthropogenic change better explained the fissioning: "there was disagreement about foraging strategies," whereby some individuals "remained natural foragers," while others became "dedicated [crop] raiders" (p.6). The promise of contemporary primatology is about joining other fields in appreciating, once again, the importance of natural history in enabling us to interpret behavior.[8]

Developing a culture of explanatory pluralism would also mean making space for positivist and interpretive ways of knowing, and hence deductive and inductive approaches, to coexist and even come together. Following the philosopher of science, Daniel Little, one of the epistemic features of science is an empirical testability criterion (1995: 42). The standard empirical approach for primatologists is to conduct quantitative behavioral observations to test hypotheses derived from theory. This method does not, however, preclude a primatologist from also attributing meaning to, for example, the relationships that develop between people and primates during the process of habituation. In this case, we could use multiple sources of evidence – participant observation, ethological observations, and ethnographic interviews – to support the attribution. This is also empiricism. It is indeed possible to be an empiricist and an interpretivist (Little 1995). The promise of contemporary primatology is therefore about taking a broader view of what science is and what we will make it moving forward.

Moving Forward

It's an energizing time to be a primatologist. Look at how far primatology has come! Back in the day, as one of my mentors expressed to me, it used to be that you could read everything that was ever written on primates. Today, various forms of social media alert me daily to the new, exciting work that has just been published. Originating from a time when there were only a handful of studies on only a handful of primates, we are now in an age in which we are able to make comparisons across the primate order and even across populations of the same species, and hence, fully appreciate the extent of behavioral variability that makes the order Primates so fascinating. Moreover, advancements in methods and technology – from minimally-invasive sampling, to remote sensor cameras, to drones, to polymerase chain reaction (PCR), to name only a few – enable us to ask and answer new questions, but also old questions in new ways. In the twenty-first century and beyond, primatologists will continue to advance the vision that Hooton and Washburn laid out for primatology. The promise of contemporary primatology that I call for in this book is neither in conflict with nor does it negate the worth of this purpose or the traditional approaches used to achieve it. Rather, it is complementary, offering insight into how and why context and contingency matters for all primate field studies. It also expands primatology's reason for being.

But, moving forward, there is more work to do. Fifteen years ago biological anthropologist, James Calcagno (2003) argued that the problem of integration is not with anthropology but with the anthropologists. I generally agree with this argument, for *we* make anthropology what it is. However, in my view, the problem is simultaneously that the institutional adoption of the four-field structure of American anthropology forces us to claim territory in one of the subfields, thereby obscuring the integrativeness of who we are and what we do as anthropologists.[9] I raise this point not to argue against upholding the four-field structure, but rather to reiterate my argument from Chapter 2 that we anthropologists can make this structure work for us, for it provides the intellectual and social space for cross-field collaborations to emerge. So, what does this mean for primatology? It means that we should strive to do more than "integration lite" primatology.[10] Primate field research that connects sociocultural variables, such as the presence or absence of human habitat disturbance, to primate behavior is important work, but we should dig deeper to explore what the disturbance is and why it occurs; we should do the anthropology to expose the context and meaning. The currently emerging scholarship on the human-primate interface illustrates a promising trend in truly integrative research in terms of both the theoretical framing and methodological approach.[11] To build on this trend we need to ensure integrative training in anthropology and allied fields and to continue pursuing intra- and interdisciplinary collaborations.

Like other mixed-methods field of inquiry, contemporary primatology also faces challenges of representation and writing. Researchers who conduct mixed-methods research must be able to communicate across entrenched divides between quantitative and qualitative research paradigms. And, indeed, readers and writers of quantitative and qualitative research belong to different "interpretive communities" who value and trust particular communication traditions (e.g., graphs and tables) over others (e.g., stories) (Sandelowski 2003: 321). The communication tradition for primatology is the standardized scientific report, which has a clear linear structure with methods separate from the results and the results separate from interpretation. For qualitative research, on the other hand, there is little standardization in reporting, as the writing up of the research is the "inquiry in the making" (Sandelowski 2003: 330). For primatology, "the heart of fieldwork" (Asquith 2010) – that is, the anecdotes and stories that capture the lives of primates and their relationships with others, including their human observers – remain relegated to popular accounts.[12] The contemporary primatology that I argue for is about making space for these important elements of primate field research in the *scholarly* work that we produce. The key will be to develop a mixed representational approach that incorporates different forms of writing and displaying our results (i.e., numerical, textual, and visual) to reach our more diverse intended audience (Greene 2008) (Figure 8.1). Identifying scholarly publication outlets that are best suited for the integrative anthropology that we do and encouraging editors to be welcoming of this type of scholarship will also be critical.[13]

Lastly, contemporary primatology traverses a complex ethical landscape (Riley & Bezanson 2018). One of the key issues is that there is a lost ethical space between what is reviewed by ethical oversight committees for human subjects research and animal subjects research (e.g., in the U.S., Institutional Review Boards (IRB) and Institutional Animal Care and Use Committees (IACUC), respectively). Consider,

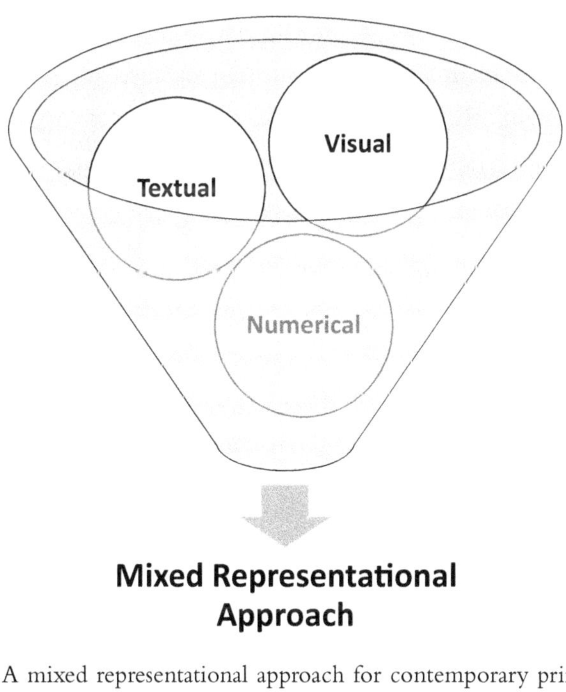

FIGURE 8.1 A mixed representational approach for contemporary primatology that incorporates diverse ways of conveying our research findings.

for example, the increasing use of remote sensor cameras in field primatology, and animal ecology, more generally. Remote sensor cameras, also called "camera traps," are extremely useful in capturing elusive behaviors of the wildlife species in question, but they can also capture human behavior. If there is no human component to the research, then there is no ethical oversight with regard to the human subjects knowingly or unknowingly involved. That said, even for research that does have a human component (e.g., ethnoprimatological research on crop feeding behavior), the guidelines of the animal subjects and human subjects protocols do not require us to report techniques used to study species not relevant to the respective review board. Using the example of camera trap research again, this means that the IACUC could clear the use of remote sensor cameras to study the primates' behavior, but the IACUC would never report to the IRB on the implications for people, nor would the researcher necessarily think to include the use of camera traps in the IRB protocol since it technically is not the tool being used to complete the research with people. This is just one example of the ethical complexity of today's primatology. To address the gaps in ethical oversight and other issues with multispecies research, some scholars have argued for the creation of a new structure of ethical oversight for primate field research; one that includes primate experts as well as scholars who are familiar with the cultural context in which the research will occur (Gruen et al. 2013). Until then, it is ultimately our responsibility to think about how the ethical considerations of our research intersect across the species we study.

Contemporary primatology that uses the anthropogenic context in which primate populations live as an opportunity to explore primate adaptability at work and how humans and primates behave together makes both conceptual and applied contributions. It contributes to our understanding of primate behavioral flexibility and how human-primate interfaces and behavior shape selection pressures. Examining the ways that humans and other primates coexist in shared ecological and social spaces not only sheds further light on what it means to be a primate, but also exposes other ways of being human. By attending to the human dimension, we also become acutely aware that efforts to conserve remaining primate populations do not take place in a vacuum. In other words, conservation is not merely an ecological problem concerned with primates and their habitats. It is also a social problem. Having a home in anthropology means that we can develop the skills to craft an approach to conservation that is integrative across the natural sciences, the social sciences, and the humanities, and hence, socially just and likely more sustainable in the long term. With contemporary primatology, we have conceptually transcended the biological-cultural divide. Moving forward, let's safeguard and promote the space to practice it.

Notes

1 It is worth pointing out that Washburn himself (with David Hamburg) noted that in "advocating the use of the evolutionary approach [for field studies of primate

behavior], we are not suggesting that this is the only method or that it is always relevant" (1968: 459).

2 For example, it is highly unlikely that ethical oversight committees today would agree with Washburn's (1983) proposition to experimentally test whether male langurs do or don't kill infant langurs. See Riley and Bezanson (2018) for a recent review on the ethics of primate fieldwork. See Janson (2012) for a discussion on how recent advances in statistical methods strengthens our use of experiments in field primatology.

3 In his retrospective piece on how the biological anthropology of today has met, and in many ways, surpassed Washburn's vision for a New Physical Anthropology, Agustín Fuentes (2010a: 4) similarly notes that by advocating for biological anthropology as a biosocial science, he may be "interpreting and expanding Washburn's position in a more radical inter- and multidisciplinary way than [Washburn] did himself."

4 In the obituary he wrote for Washburn, Jonathan Marks (2000) notes that in advancing his "New Physical Anthropology," Washburn may have overemphasized the importance of evolutionary biology to the neglect of the anthropology that gave it form in the first place, which in turn may have contributed to the biological-sociocultural divide discussed in Chapter 2.

5 Washburn (1968) argued that academic institutions should be organized around research problems rather than by disciplines/departments. Like many of his ideas, this is a prescient one that in fact has been implemented at one of the nation's top research universities, Arizona State University (ASU). Under the vision of President Michael Crow, much of ASU's faculty is part of schools and is organized into "faculties" rather than departments, thereby encouraging interdisciplinary exchange and collaboration.

6 Or for that matter, ethologial observations alone: "Data collection does not end when the gorillas bed down for the night. As you walk back to camp you hear the sounds, experience the forest with people, eat, talk about the day, and realize the ways in which life in a community is integral to your understanding of the relationship to and between people, primates, the environment, and larger structural change" (Remis & Robinson 2017: 192).

7 See Goodman and Leatherman (1998), a follow-up by Goodman (2013), and a reply to Goodman (2013) by Calcagno (2014) who argues that biological anthropologists' goal is not so much bringing culture back into biology, and biology back in anthropology, but keeping them both together (as did Boas) but perhaps with increased emphasis and for exploring new directions.

8 Here I am referring to behavioral ecology and ecology. See Altmann and Altmann (2003), Gordon (2011).

9 Even though my job description was listed as a primate behavior and conservation, I was hired as one of the "biological anthropologists." Accordingly, I am (happily) responsible for teaching the biological anthropology courses, but I would not restrict my research interests to biological anthropology, for as I relay in the introduction, I am a hybrid.

10 Here I am borrowing the term Alan Goodman (2013, 2014) used to describe biocultural research.

11 See Waller (2016), Dore et al. (2017), special issue in *International Journal of Primatology* (vol.39, issue 5) on "Ethnoprimatology in the 21st Century," and Behie et al. (2019).

12 See science studies scholar, Amanda Rees' fascinating work on primate field sites and the making of authentic observations (2006) and popularized accounts of primatology (2007).

13 I was asked after a talk at a scholarly conference on whether we should create a journal specifically devoted to ethnoprimatology. I have mixed feelings about this; on the one hand, it would be an extremely fruitful place to reach fellow primatologists interested

in the human-primate interface and the kind of intra- and interdisciplinary primatology that I am calling for. On the other hand, how would we then reach the broader scholarly community doing integrative anthropology, more generally, and working on human-wildlife relations, more specifically? Also, how would we reach our skeptics? The good news is that a number of the mainstream primatology journals (e.g., the *International Journal of Primatology* and *Folia Primatologica*), are now more open to the mixed representational approach I describe.

References

Altmann SA, Altmann J. 2003. The transformation of behaviour field studies. *Animal Behaviour* 65: 413–23.

Armelagos GJ, Gerven DPV. 2003. A century of skeletal biology and paleopathology: contrasts, contradictions, and conflicts. *American Anthropologist* 105: 53–64.

Ashley-Montagu MF. 1940. A cursory examination of the relations between physical and social anthropology. *American Journal of Physical Anthropology* 26: 41–61.

Asquith PJ. 2010. Natural homes: primate fieldwork and the anthropological method. In *Centralizing Fieldwork: Critical Perspectives from Primatology, Biological Anthropology and Social Anthropology*, ed. J MacClancy, A Fuentes, pp. 242–55. Oxford: Berghahn.

Behie AM, Teichroeb JA, Malone N. 2019. *Primate Research and Conservation in the Anthropocene*. Cambridge: Cambridge University Press.

Berger-Tal O, Polak T, Oron A, Lubin Y, Kotler BP, Saltz D. 2011. Integrating animal behavior and conservation biology: a conceptual framework. *Behavioral Ecology* 22: 236–39.

Bicca-Marques JC, Garber PA. 2004. Use of spatial, visual, and olfactory information during foraging in wild nocturnal and diurnal anthropoids: A field experiment comparing *Aotus, Callicebus*, and *Saguinus*. *American Journal of Primatology* 62: 171–87.

Calcagno JM. 2003. Keeping biological anthropology in anthropology, and anthropology in biology. *American Anthropologist* 105: 6–15.

Calcagno JM. 2014. Still keeping cultural and biological anthropology together: A response to Alan Goodman's AAA Presidential Address. *American Anthropologist* 116: 404–05.

Cheney DL, Seyfarth RM. 1982. How vervet monkeys perceive their grunts: Field playback experiments. *Animal Behaviour* 30: 739–51.

Dore KM, Riley EP, Fuentes A. 2017. *Ethnoprimatology: A Practical Guide to Research at the Human–Nonhuman Primate Interface*. Cambridge: Cambridge University Press.

Fuentes A. 2010a. The new biological anthropology: Bringing Washburn's New Physical Anthropology into 2010 and beyond–The 2008 AAPA luncheon lecture. *Yearbook of Physical Anthropology* 53: 2–12.

Fuentes A. 2010b. Measuring meaning and understanding in primatological and biological anthropology fieldwork: context and practice. In *Centralizing Fieldwork: Critical Perspectives from Biological and Social Anthropology*, ed. J MacClancy, A Fuentes, pp. 121–36. New York: Berghan Books.

Garber PA, Gomes DF, Bicca-Marques JC. 2012. Experimental field study of problem-solving using tools in free-ranging capuchins (*Sapajus nigritus,* formerly *Cebus nigritus*). *American Journal of Primatology* 74: 344–58.

Goodman AH. 2013. Bringing culture into human biology and biology back into anthropology. *American Anthropologist* 115: 359–73.

Goodman AH, Leatherman TL, eds. 1998. *Building a New Biocultural Synthesis: Political-Economic Perspectives on Human Biology*. Ann Arbor, MI: University of Michigan Press.

Gordon DM. 2011. The fusion of behavioral ecology and ecology. *Behavioral Ecology* 22: 225–30.

Greene JC. 2008. Is mixed methods social inquiry a distinctive methodology? *Journal of Mixed Methods Research* 2: 7–22.

Gruen L, Fultz A, Pruetz J. 2013. Ethical issues in African great ape field studies. *ILAR Journal* 54: 24–32.

Gumert MD, Fuentes A, Engel G, Jones-Engel L. 2011. Future directions for research and conservation of long-tailed macaque populations. In *Monkeys on the Edge: Ecology and Management of Long-tailed Macaques and their Interface with Humans*, ed. MD Gumert, A Fuentes, L Jones-Engel, pp. 328–53. Cambridge: Cambridge University Press.

Ingold T. 2007. The trouble with 'evolutionary biology'. *Anthropology Today* 23: 13–17.

Janson CH. 2012. Reconciling rigor and range: Observations, experiments, and quasi-experiments in field primatology. *International Journal of Primatology* 33: 520–41.

Leca JB, Gunst N, Huffman MA. 2010. Indirect social influence in the maintenance of the stone-handling tradition in Japanese macaques, *Macaca fuscata*. *Animal Behaviour* 79: 117–26.

Little D. 1995. Objectivity, truth and method: A philosopher's perspective on the social sciences. *Anthropology News* 7: 42–43.

Marks J. 2000. Sherwood Washburn, 1911–2000. *Evolutionary Anthropology: Issues, News, and Reviews* 9: 225–26.

Rees A. 2006. A place that answers questions: rimatological field sites and the making of authentic observations. *Studies in History and Philosophy of Biological and Biomedical Sciences* 37: 311–33.

Rees A. 2007. Reflections on the field: Primatology, popular science, and the politics of personhood. *Social Studies of Science* 37: 881–907.

Remis MJ, Jost Robinson CA. 2017. Nonhuman primates and "others" in the Dzanga Sangha Reserve: The role of anthropology and multispecies approaches in ethnoprimatology. In *Ethnoprimatology: A Practical Guide to Research at the Human-Primate Interface*, ed. KM Dore, EP Riley, A Fuentes, pp. 190–205. Cambridge: Cambridge University Press.

Riley EP, Bezanson M. 2018. Ethics of primate fieldwork: Toward an ethically-engaged primatology. *Annual Review of Anthropology*.

Sandelowski M. 2003. Tables or tableaux?: The challenges of writing and reading mixed methods studies. In *Handbook of Mixed Methods in Social & Behavioral Research*, ed. A Tashakkori, C Teddlie, pp. 321–50. London: Sage.

Strier KB. 2009. Seeing the forest through the seeds: Mechanisms of primate behavioral diversity from individuals to populations and beyond. *Current Anthropology* 50: 213–28.

Strum SC. 2012. Darwin's monkey: Why baboons can't become human. *American Journal of Physical Anthropology* 149: 3–23.

Waller MT. 2016. *Ethnoprimatology: Primate conservation in the 21st Century*. New York: Springer.

Washburn SL. 1968. One hundred years of biological anthropology. In *One Hundred Years of Anthropology*, ed. JO Brew, pp. 97–115. Cambridge, MA: Harvard University Press.

Washburn SL. 1973. The promise of primatology. *American Journal of Physical Anthropology* 38: 177–82.

Washburn SL. 1983. Evolution of a teacher. *Annual Review of Anthropology* 12: 1–24.

Washburn SL, Hamburg DA. 1968. Aggressive behavior in Old World monkeys and apes. In *Primates: Studies in Adaptation and Variability*, ed. PC Jay, pp. 458–78. New York: Holt, Rinehardt & Winston.

INDEX

Note: Page numbers in *italic* denote figures and in **bold** denote tables, end of chapter notes are denoted by a letter n between page number and note number.

CPSIA information can be obtained
at www.ICGtesting.com
Printed in the USA
BVHW040729221119
564306BV00016B/46/P

9 781629 580708